THE EARTH

THE EARTH

Discovering our place in the universe

Josefina Arce
Scientist, Educator, and Farmer

This book was written and designed by Josefina Arce in Villalba, Puerto Rico, as a celebration of life, science, and nature.

Title: The Earth: Discovering Our Place in the Universe
Author: Josefina Arce
Visual Artist: Josefina Arce
Design and Layout: Josefina Arce
First Edition: 2025
ISBN: 978-0-9725632-8-4
Printed in the United States of America

This book was printed on demand, reducing waste and contributing to a more sustainable publishing model.

Dedication / In Memoriam

To Dr. Jane Goodall

— friend of The Earth and teacher of humanity —

Today I pay tribute to you with gratitude and tenderness.
You showed us that chimpanzees also feel, love, laugh, and cry;
that we are not so different, and that life is woven from the same
threads in all beings who dwell on The Earth.
You taught us that by giving an animal a name, we recognize it as
someone, not *something*.
You reminded us that what we call "wildlife" also knows bonds,
emotions, and awareness.

Your example—always calm yet steadfast—accompanied me in the
creation of this book.
It was your voice—loving and urgent at once—that inspired me to
write about our place in the universe and about the need to care for
The Earth that sustains us.

Thank you for your life and legacy, which will continue to illuminate
the way forward.

With affection and respect,

Josefina Arce Quiñones
Puerto Rico, October 2025

Author's Note

This book was born from a question both simple and immense: How did we come to be here, on this planet, sharing life with so many different and fragile forms?

As a scientist, educator, and farmer, I've learned that The Earth speaks to us with many voices: the water that runs, the wind that caresses, the soil that holds us up, and every being that is born and dies in this continuous cycle. Listening to those voices has led me to recognize that our place in the universe is measured not only in distances or years, but in the responsibility we bear to care for what sustains life.

I wrote these pages with children, young people, and families in mind—those who seek to understand, with wonder and clarity, how science and poetry intertwine to tell the story of our home.

May this book be an invitation to look at the sky with questions, at The Earth with gratitude, and at ourselves with the certainty that we are part of a whole—larger, older, and more alive than we usually imagine.

I also hope that, as we turn these pages, we recognize the privilege of being alive and the responsibility we carry, as conscious beings, to keep our home in order. May we choose a healthier life, in harmony with nature and shared joy. For this book is not only to admire what already exists, but to remind us that we are called to care for it and to act—each from where we are and exploring every possibility. And to act in truth: The Earth responds to what is real, not to what we wish to believe. Caring for it requires distinguishing evidence from propaganda, and learning to think with rigor and with love.

— Josefina Arce Quiñones

CONTENTS

Note To The Reader

This book reads like a story and is meant to be savored slowly. It's a narrative that invites you to look at The Earth with wonder and respect.

Beyond the poetic voice and the reflections, you'll find resources to support you at different levels of reading:

Letters and Final Reflections: after Parts I–IV you'll find two letters—one addressed to those who believe in Creation and another to the scientific and educational community—followed by a reflection titled What Truly Matters. All are words written for those who wish to listen and to act.

Glossary: at the end you'll find simple explanations of some terms that appear in the narrative of Parts I–IV. If a term is new to you, you can look it up there and learn it at your own pace.

Complementary References: selected to help you deepen into the topics covered. They are not requirements but invitations—use them to explore further and consider additional perspectives.

Part V — Voices for Action: this part is not meant to be read straight through, but to be consulted and revisited like a toolbox:
The Earth's Mirror: Evidence and Data gathers the charts and scientific references that support what is told in the earlier parts. Paths That Already Exist offers current examples and traditional trades that show practical ways to turn inspiration into action.

Acknowledgments: because every word stands on the work of those who researched, taught, and shared—and because this book is also the fruit of many hands.

You can read it from beginning to end as a story that unfolds before you, or move between the narrative and these resources whenever you need. The intention is not only that you enjoy the reading, but also that you find clarifications, scientific grounding, and concrete alternatives for action.

Before you begin...

Close your eyes for a moment.
Breathe.
Feel the air moving in and out.
Your body is alive: every cell breathes with you.

The story you're about to read is alive, too.
It's not just knowledge; it's a call to remember.

Remember that your body comes from The Earth:
from minerals, from water, from the sun, and from the life that
existed before you.
Everything that breathes is with you in this moment.

Read slowly.
Listen to the words as if they were wind, as if they were root.
Let your body feel before your mind analyzes.

Because understanding The Earth is also
remembering that we are Earth.

Introduction

I am a mother, grandmother, scientist, farmer, and Professor Emerita of Chemistry at the University of Puerto Rico. My life has always been woven with science, education, and The Earth itself. I earned a Ph.D. in organic chemistry, taught at the university for more than 37 years, and led large-scale educational projects that helped transform the way science and mathematics are taught in our schools and universities.

That academic path gave me knowledge, but it also left me with a deeper truth: life cannot be separated into disciplines, as we often teach it. Human beings and nature can only be understood together. My own story—from a little girl planting seeds on the balconies of Old San Juan, Puerto Rico, to seventeen years as a farmer on my land—reminds me that this connection is both real and vital.

This book is born from a deep love for our Mother Earth (Atabey, in the Taíno language). I don't wish to write a long or heavy volume, but rather a clear and condensed guide filled with essential ideas—something you can read in a short time, return to often, and always find something new. It's written in simple language, with key concepts from our human story, yet with the hope that it will inspire you to keep learning and searching.

I'm not here to scold or to frighten. I'm here to invite you to look differently—to notice the thin layer where all life unfolds, to recognize how we have harmed it, and to rediscover how we can care for it. I speak as a scientist, an educator, and a farmer—but also as one of you, with the same vulnerability and the same longing to live on a healthy planet.

What moves me to write is the conviction that we can still set our compass straight and place first what should have always come first: **life itself.**

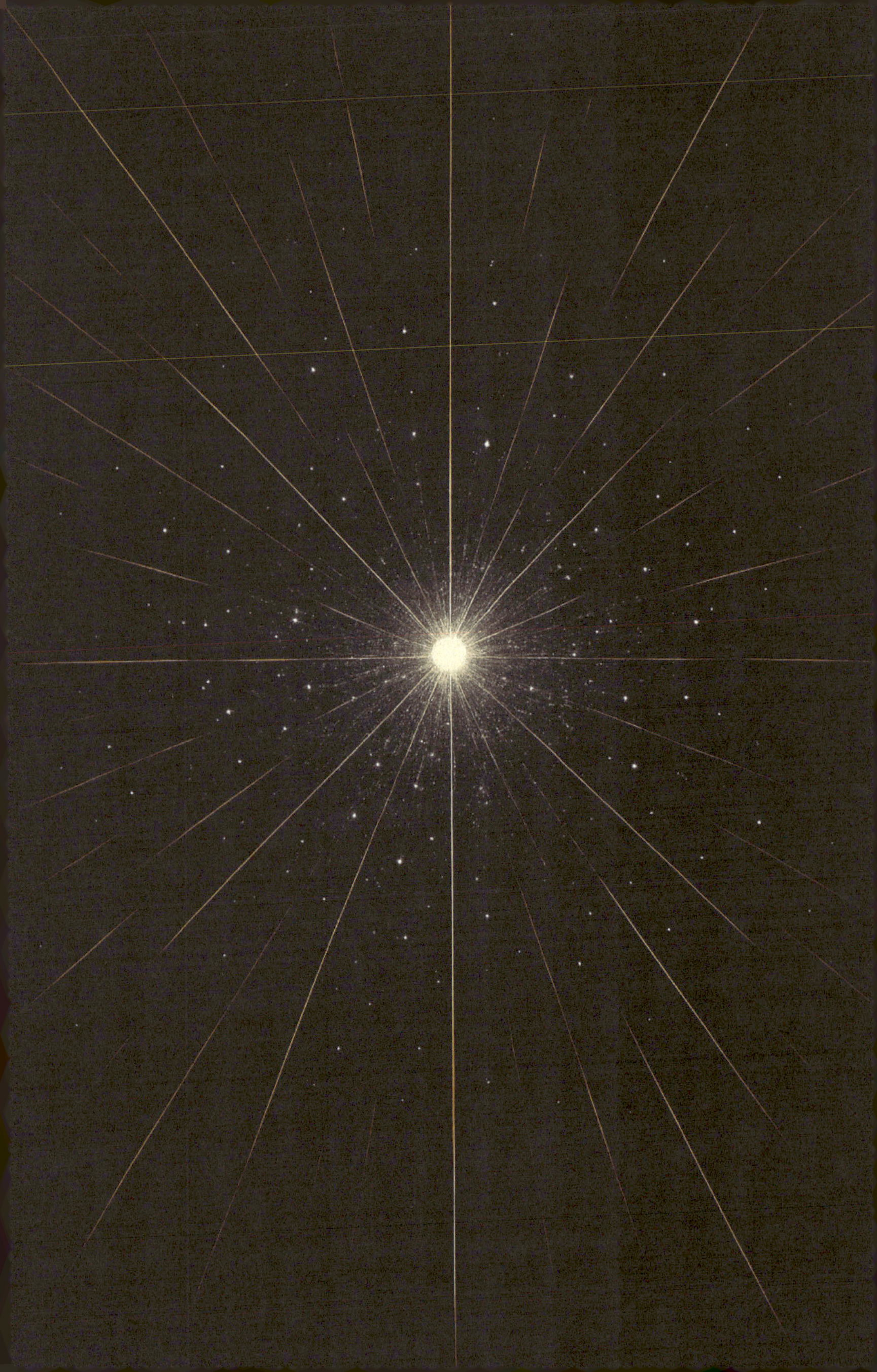

PART I

THE COSMIC STAGE

"The cosmos is within us.
We are made of star stuff.
We are a way for the cosmos to know itself."

— Carl Sagan, astronomer and science communicator

Chapter 1

A Beginning Of Stars

Nothing.

Thirteen point eight billion years ago, there was nothing our senses could recognize—no time, no space, no matter. Only a mystery compressed into a point smaller than a grain of sand.

What existed before that? We don't know. And it's important to say so: science cannot speak of what cannot be observed or measured. We can only reach as far as our instruments and theories allow. That beginning is what we call the Big Bang. And it isn't just a wild idea—it's the explanation that best fits everything we see in the cosmos today.

Then it happened.

In a fraction of a second, that point burst into light, energy, and expansion. The Big Bang wasn't an explosion in space. It was the creation of space itself.

First, there was pure energy.

Then, tiny particles: protons, neutrons, electrons.

Within minutes, they joined to form the first nuclei—hydrogen and helium, the elemental building blocks of everything that would come after.

For hundreds of millions of years, darkness ruled. The universe expanded and cooled, but there were still no stars. Space was like a vast, cold ocean—silent and nearly empty. Even today, most of the cosmos remains that way: a dark stillness, almost without atoms, without air, without molecules. Only here and there, a spark of light appears.

Until, under the pull of gravity, immense clouds of hydrogen collapsed and ignited. Thus the first stars were born—nuclear furnaces that transformed simple matter into complex elements. The oxygen you breathe, the calcium in your bones, the iron in your blood—all were forged in the fiery hearts of stars that died long before Earth existed. Every atom in your body is recycled stardust, a traveler in a universe that never stopped transforming. When you look at a jewel, a glass of water, or your own hand, you are seeing bits of ancient stars still journeying within you.

Then came galaxies—millions of suns, planets, and moons. Among that infinite ocean, one small solar system began to orbit a white star: the Sun.

And on the third planet from that star, the right conditions appeared for the improbable to happen—life.

Here's something fascinating: the natural laws tell us that the universe tends toward disorder. Energy disperses, structures break apart, everything moves toward maximum entropy. Yet under certain circumstances, islands of order can emerge in the midst of chaos. That's how stars formed, how planets formed, and—much later—how life began. The laws aren't broken; for something as organized as a living being to exist, the rest of the universe must increase its disorder even more. It's as if, in the middle of an infinite noise, an unexpected melody was born.

When we look at the night sky today, we're looking at an ancient map. The light of the stars takes years, centuries, or even millennia to reach our eyes. It's like receiving letters from the cosmic past. And in that vast spectacle, we must remember something essential: we are part of that story.

If the history of the universe were compressed into a single calendar year, humanity would appear in the last seconds of December 31. Our existence is just a blink in the scale of the cosmos—yet that blink is full of meaning.

Humanity is not an isolated accident in some corner of the universe. We are the outcome of a process that began with a spark of energy, continued with the dance of particles, and found its echo in every leaf, every river, every bird's song.

Imagine those first atoms as scattered letters. Over time, they joined to form words, sentences, chapters, and finally whole books of complexity. Or like loose musical notes that, when brought together, became melodies—and later, a great moving symphony. Life is part of that music, and we belong to that score.

Understanding our cosmic origin does not distance us from Earth. On the contrary, it reminds us that everything is interwoven. That on this unique, improbable, living planet, billions of years of history have come together, waiting to be honored.

We are the most recent echo of a story that is still being written.

Chapter 2

Earth: An Improbable Oasis

Among billions of stars, countless planets spin—worlds of rock, ice, and fire. Most of them are barren, too cold, too hot, too dry, or too acidic. Across the vastness of the universe, silence reigns.

And yet—here we are.

On this blue dot suspended in darkness, life took root and blossomed.

Earth is not the largest planet, nor the brightest. It isn't the center of anything. And still, it brings together such a perfect balance of conditions that it feels like an almost impossible miracle:

- **Just the right distance from the Sun.** A little closer, and we'd be a furnace like Venus; a little farther, a frozen desert like Mars. A tiny difference, in cosmic terms, decides between fire, ice... or life.
- **An almost circular orbit and a tilted axis.** Together they give us mild seasons and repeating cycles with a steady rhythm.
- **A protective atmosphere.** A precise blend of nitrogen, oxygen, and other gases filters deadly radiation and holds just enough warmth for life to thrive.
- **Liquid water.** Not only in vast oceans, but in rivers, lakes, clouds—and in every living cell.
- **An invisible magnetic field.** A shield against solar winds that would otherwise strip the planet of its atmosphere.
- **A faithful moon.** Its gravity pulls the tides and keeps Earth's axis steady, preventing chaotic spins that would destroy climate stability.

Each of these conditions, on its own, is rare in the universe. Together, working in harmony, they made Earth a true cosmic oasis.

It's as if Earth had won the cosmic lottery.

But instead of money, the prize was breathable air and liquid water flowing freely across its surface and beneath it.

It wasn't mere chance—it was the outcome of physical, chemical, and astronomical processes woven together over 4.5 billion years. And in that precise setting, molecules began to organize, to copy themselves, to transform... until they could sing and tell stories.

Today we breathe without thinking about the exact mix of gases that enters our lungs.

We walk without noticing that beneath our feet, a magnetic field shields us from invisible storms.

We watch the tides rise and fall, unaware that they are the pulse of the Moon caressing the ocean.

We live as if all this were obvious.

But it isn't.

It is highly improbable.

It is fragile.

And that is precisely why it is **sacred**.

To understand Earth as an improbable oasis is not just an interesting fact.

It is an urgent reminder: there is no other place where we can live. There is no spare planet waiting in silence.

Our only home is here—alive, fragile, and beating beneath our feet.

"Our planet is unique—a perfect balance of distance from the Sun, liquid water, and a protective atmosphere. A miracle, truly."

— Sir David Attenborough, naturalist and documentarian

Chapter 3

The Miracle Of Water

Across the universe there are oceans of methane, like on Titan, acid storms on Venus, and seas of ammonia on distant worlds. But only here, on Earth, does water sing—as river, cloud, tear, and sea.

Water seems simple enough: one oxygen atom embraced by two of hydrogen. A light, tiny molecule. And yet, from that simplicity arises a miracle of properties that sustain all life on Earth.

- **The heat it stores and releases.**
  Water warms slowly and cools with patience. Because of that, oceans and lakes soften the planet's climate, preventing Earth from becoming a blazing desert by day and a frozen wasteland by night. And when water absorbs too much heat, it gives it back as storms and hurricanes—solar energy transformed into wind and rain.

- **The ice that floats.**
  Unlike almost any other substance, solid water is less dense than its liquid form. As it freezes, its molecules move apart, taking up more space—and so, ice floats. Lakes and rivers freeze from the top down, keeping their depths liquid and allowing fish and plants to survive beneath a frozen roof. If it weren't that way, aquatic life would vanish each winter across much of Earth.

- **The cycle that never rests.**
  Water evaporates and rises to the sky, condenses into clouds, returns as rain, runs through rivers, hides within the ground, and begins again.
  An endless cycle that moves the sap of forests, renews the soil, and refreshes the air.

- **The absent heart of trees.**
  Water climbs through stems and trunks thanks to cohesion and capillary action—droplets holding their molecules together, allowing them to rise through narrow tubes. Thus it

ascends without pumps or hearts, feeding leaves even thirty meters above the ground.

- **The mirror of life.**
  Water is the best solvent known. In it travel salts, sugars, oxygen, and nutrients. Every cell is, in essence, a tiny ocean—a little pouch of water where the molecules of life float. Blood, sap, tears, and sweat are variations of that same primordial sea.

- **The light that passes through.**
  Transparent, it lets sunlight reach rivers and oceans. Thanks to that, phytoplankton and algae perform photosynthesis beneath the surface, producing much of the oxygen we breathe.

Water exists throughout the universe as ice or vapor, but only on Earth does it flow so abundantly in liquid form—balanced with its other phases. And only here did that balance become the cradle of life.

That's why every drop we drink, every passing cloud, every wave breaking on the shore, is part of the same miracle—a continuous one that reminds us water cannot be treated as just another resource.

Every time we open a faucet and see clear water run, we are witnessing one of the planet's greatest miracles—so common that we forget to marvel.

What would your day look like if water didn't reach your tap tomorrow?

Our entire routine would collapse.

Water is Earth's invisible sap, and caring for it must be our highest duty.

Interlude — Bridge To Part II

This thin layer of the planet—made of air, water, and soil—is the stage where the first spark of life appeared. Within this narrow band, the first cells emerged, and here they continue to renew themselves every day.

From this point, we leave behind the vast and distant to draw closer. We'll look at how, in this small stage, the inert began to pulse with life. Like a note finally finding its rhythm, matter began to transform into living forms— and with them, the story of life began, a story still being written today.

"It is not the strongest of the species that survives,
nor the most intelligent,
but the one most adaptable to change."

— Charles Darwin, 1859, On the Origin of Species

PART II

THE AWAKENING OF LIFE

Chapter 4

The Layer Of Life

The entire story of life takes place within a tiny space—a thin layer stretching only a few miles up into the atmosphere and a few miles down into Earth's crust.

This space makes up less than one percent of the planet's volume.

Outside this layer, there are no birds, no rivers, no forests. Above it, the air becomes too thin to breathe, and radiation burns without mercy. Below it, heat and pressure make life impossible.

It is within this narrow margin—barely a living membrane around the planet—where everything happens:

- Where the Sun delivers its energy, and plants transform it into food.
- Where water moves through its endless cycle of vapor, clouds, rain, rivers, and seas.
- Where nutrients travel from soil to roots, and from there into every food chain.
- Where the living and the nonliving meet in dialogue: minerals, air, microorganisms, animals, and human beings—all intertwined.

The balance that sustains this layer depends on the cooperation of every part:

the microscopic algae in the sea releasing oxygen,

the worm that stirs the soil and lets the roots breathe,

the tree that captures carbon and cools the air,

the predator that keeps its prey in check so the forest doesn't collapse.

In this living network, opinions do not change outcomes: if one thread is broken, the whole web feels it.
Biology does not negotiate with stories; it answers to causes and consequences.

It's like living in a house with glass walls—it seems safe, but one blow can shatter it.

When a river becomes polluted in your town, or when a summer arrives hotter than ever, what's at risk is that thin layer that sustains us all.

Earth will go on existing, but the orchestra falls out of tune—and with it, our own survival is at stake.

To understand that we live within this delicate layer should awaken reverence.

We are **privileged** inhabitants of a tiny space in the cosmos where life is possible.

We know of no other place like it.

Protecting this layer is not an act of compassion toward nature.
It is an act of self-love, of survival, of gratitude—
because caring for it means caring for ourselves, for our descendants, and for everything we love.

All life unfolds within this thin layer on a planet in a vast universe.
If it breaks, there is no replacement.

"The living environment is the biosphere—the thin layer around the world of living organisms. We are part of it. Our existence depends on it in ways people have yet to begin to appreciate."

— E. O. Wilson, American biologist

Chapter 5

The Spark Of Life

There was a moment— an improbable flash in the vast history of the universe.

Within the thin layer of air, water, and soil that wraps around Earth, conditions emerged that no nearby planet could sustain: the right temperature, the right pressure, the right chemical ingredients, and a source of energy that set them in motion.

It was a precise point in time and space where inert matter, against all odds, began a flow of energy that has never stopped.

That moment was the spark— what the faithful might call Creation, and what science describes as the beginning of a chain of exquisitely synchronized reactions.

Life runs on a river of electrons— a dance of molecules seeking and binding, reaction after reaction, links in an invisible chain.

It began there, where rock, water, and air met, and a flicker of chemical or physical energy learned to sustain itself.

Since then, the current has never ceased: each cell is a small vortex that shelters the fire, each organism a network of reactions holding one another in balance.

When the elements are missing, when the river dries up and the electrons stop flowing, the chain is broken—and we say that life goes out. But as long as the chain continues, as long as the molecules keep dancing in fragile equilibrium, the spark endures. And the universe keeps singing, creating millions of living forms.

Somewhere in the ancient ocean, chemistry became cradle. Alkaline hydrothermal vents offered porous mineral walls and natural proton gradients—tiny batteries capable of driving reactions. There, matter began capturing energy and organizing itself into cycles.

The traits associated with the Last Universal Common Ancestor— LUCA—point to such settings: environments rich in H_2, CO_2, and iron, consistent with those hydrothermal vents.

LUCA was likely more a metabolic ancestor than an anatomical one—
a network of reactions rather than a single-shaped being.

In those early times, life also shared genes laterally: a constant exchange
that sped up innovation and blurred boundaries.

From that first organization arose millions of lineages—
some brief as a sigh, others resilient enough to evolve and endure for
billions of years.

Every living being today, from the worm to the human,
is heir to that first thread woven into the fabric of time.

Life is knowledge in chemical form:

it knows how to replicate,

how to repair,

how to persist.

Life did not begin perfect.

It was a river of experiments—some currents vanished, others kept
flowing.

Each new form was a test in Earth's silent workshop.
Some flourished; others faded.

And so, step by step, the long story of the living was written.

Chapter 6

The Great Transformation

For millions of years, the first organisms lived in a world without oxygen.

To them, that gas was poison.

Then something astonishing happened: tiny living beings discovered how to use sunlight to make the food that stores the chemical energy needed to sustain life.

Every leaf we see today descends from that ancient miracle—an ancestral solar panel. As these organisms captured energy, they released oxygen as waste. What once seemed toxic became the foundation of a new atmosphere.

Our oldest records of photosynthetic communities appear in stromatolites—layered structures built by microbes—and in fossils of cyanobacteria from the early Proterozoic.

This planetary reconfiguration left marks. About 2.4 billion years ago, the Great Oxidation Event transformed the air and the oceans. Before that, the oxygen produced had been swallowed by dissolved iron and other sinks. When those sinks became saturated, O_2 began to accumulate, leaving in the rocks the distinctive bands of iron that we now read as evidence of that turning point.

At the same time as they released oxygen, these organisms were pulling carbon from the air as CO_2 and incorporating it into their own organic matter—solar radiant energy transformed into potential energy stored in chemical bonds.

Layer after layer, for millions of years, the remains of that life became buried beneath sediments and ancient waters, storing that chemical energy in reservoirs that, over geological time, would take solid, liquid, and gaseous forms.

Without plan or intention, life gradually removed carbon from the atmosphere—carbon that once existed as CO_2—reducing the greenhouse effect and helping the planet cool. That thermal pause cleared the way for aerobic organisms to flourish and populate The Earth.

In time, part of that oxygen formed ozone (O_3) high in the atmosphere—a shield against ultraviolet radiation that would later make life on land possible.

That innovation was the alliance between sunlight and life—an ancient pact with the white star that sustains us.

The air began to fill with oxygen, the sky changed its chemistry, the seas transformed.

Without meaning to, those green bacteria created the conditions that, much later, would allow us to breathe.

Oxygen, however, was a double-edged sword.

With it came the ability to release far more energy—but also the risk of burning from within. Organisms had to invent defenses, strategies, and balance. Life learned to use as resource what had once seemed danger.

From then on, the rule became clear:
use the energy of the present, close the cycles.
That is the natural law that allows life; anything that defies it will bring consequences, no matter what story we tell ourselves.

Out of that transformation arose new forms—more complex, more diverse. Plants, fungi, and animals are heirs of that great turning: the moment when the planet turned green and the air became breathable.

Every breakfast we eat begins with the Sun.

The energy that ripens the fruit and grows the coffee is the same that, billions of years ago, became sap through that ancient pact between light and life.

That invisible gas, once poison, is now the breath that sustains our lives.

$$6\ CO_2 + 6\ H_2O \rightarrow C_6H_{12}O_6 + 6\ O_2$$

CARBON DIOXIDE   WATER   GLUCOSE   OXYGEN

Chapter 7

The Web Of Evolution

Life did not stay still.

Once it was ignited, it began to transform without rest— a workshop of forms, where some lasted only geological instants, and others endured for entire eras.

Sometimes the path favored strength, magnitude, and power. But a giant without support eventually falls. The enormous creatures that once ruled the seas and skies learned that lesson when balance was broken and they vanished.

The true key to evolution was not strength or cleverness alone, but the ability to integrate. Life is a fabric of mutual favors— bacteria living inside cells, fungi feeding roots, animals dispersing seeds.

A web of harmonious relationships, like countless collaborating groups. It has always been a dance between the tiny and the vast, where the invisible sustains the visible.

Cooperation reached even the heart of the cell: endosymbiosis. An ancient partnership among microbes gave rise to mitochondria, and later, in plants, to chloroplasts— that's why these organelles still carry bacterial traits. That internal alliance multiplied the energy available and opened the door to more complex forms of life.

Mitochondria made it possible to extract, with far greater efficiency, the potential energy stored in organic compounds—a legacy of photosynthesis, in which life had already learned to transform sunlight into chemical bonds.

Much later, the oceans became home to soft-bodied creatures of the Ediacaran period, and about 541 million years ago, the Cambrian recorded an explosion of new body plans—the Cambrian Explosion, a burst of diversity that reshaped life on Earth.

Thus life moved forward in a spiral—from microscopic to gigantic, from simple to complex.

Through millions of years, Earth was quietly preparing for something unprecedented: the emergence of beings capable of contemplating their own story.

When plants first ventured onto land, they did not go alone. They were accompanied by fungi. Mycorrhizae—the union of root and fungus—improved nutrition, enhanced water management, and made possible the first terrestrial ecosystems.

Every step in evolution was also an act of cooperation. What we are today is the fruit of millions of silent alliances. And every failure—every branch that went extinct—was also a lesson etched into the planet's memory.

No living being survives alone.

Each inherits the world that allows it to exist.

Introduction To Part III

Life has never walked alone.

Every living being is sustained by others: a tree breathes because invisible fungi feed it from below, a cloud forms because millions of leaves release vapor, an animal survives because plants turn sunlight into food.

Everything is woven into a deep network—sometimes hidden, yet indispensable—the web that holds up everything we see, from forests and rivers to our own bodies.

In this part, we'll enter that web. We'll see how matter moves through cycles that never stop, how the tiny supports the immense, and how the vast diversity of life pulses within a single shared story.

Each year, the land and the ocean together absorb roughly half of the CO_2 we emit. They are the planet's invisible brakes—but signs already show they are weakening under global warming,making it ever more urgent to care for the web that sustains them.

What began as isolated molecules has become an immense network. Now we step into that living fabric that sustains us all.

PART III

THE WEB OF LIFE

"Everything spins, rotates, vibrates, repeats, pulses, and resonates—circles of life born from the pulse of light, breathing as they expand in a spiral toward eternity."

— Suzy Kassem, poet and philosopher

Chapter 8

The Dance Of Matter

Evolution filled The Earth with diversity—creatures large and small, plants, fungi, animals. But none of them could exist on their own.

From the beginning, life discovered that it could only endure through a greater balance: the constant recycling of matter. Scientists call this web of relationships an ecosystem.

Every atom you breathe, every drop you drink, every bite that nourishes you has passed through countless other bodies before reaching yours. The carbon in your blood may once have flowed through a tree's sap; the oxygen entering your lungs was released by microalgae in the sea; the water that quenches your thirst was once a cloud, a raindrop, a river, a deep ocean current—before becoming part of you.

That's why understanding cause and effect is an act of care.
When we mistake desire for data, we act blindly—
and harm the very systems that sustain us.

Earth is not just rock, ocean, and air.
It is a living system of subtle exchanges:

- •Forests absorb carbon dioxide and give back oxygen.
- •Oceans regulate temperature and give birth to clouds.
- •Soil stores water and nutrients that feed plants.
- •Invisible microorganisms transform matter and decide whether a seed will thrive or not.

When air masses pass over forests, they carry more moisture—and it rains more downstream than it would over deforested land. Trees do more than capture carbon: they move water and cool entire regions.

By drawing CO_2 from the atmosphere and weaving it into living matter, ecosystems keep the carbon cycle in balance, moderate the planet's heat, and sustain the conditions that allow life to flourish.

Nothing exists in isolation.

When a species disappears, a thread in the web snaps—and the whole system feels the loss.

Yet for millions of years, the planet self-regulated with remarkable precision: ice and warmth, droughts and rain, life and death—all flowing through cycles that balanced and nourished one another.

The dance of matter is Earth's oldest heartbeat.

To understand it is the first step toward learning to **respect** it.

"If we want clean water, we have to return biology to our soils. If we want to grow and harvest food, we have to build soil and fertility over time, not destroy them."

— Elaine Ingham, soil microbiologist

Chapter 9

The Invisible Fabric

When we look at a forest, we see trunks, leaves, and birds.

When we look at a river, we see the gleaming water that runs through it. But rarely do we look beneath our feet. There, hidden from sight, pulses a tiny universe that sustains everything visible.

A single gram of soil can contain nearly a billion bacteria—along with fungi, archaea, and microscopic animals. We don't see that bustling world, but within it, fertility is decided and the nutrient cycles unfold.

A handful of soil holds more living organisms than there are stars in our galaxy. Bacteria and fungi turn the dead into nourishment. Earthworms and insects aerate the soil, making it fertile. Invisible roots intertwine with fungi in underground networks that connect trees to one another, allowing them to share nutrients and signals.

Recent studies estimate that plants transfer about 13 gigatons of CO_2 equivalent per year to mycorrhizal fungi—a massive underground flow that helps explain why soils store so much carbon. Scientists are still investigating how much of that carbon remains stabilized over the long term.

Soil is not just dirt.

It is a living community, a silent laboratory that sustains the surface of life.

- A tall tree can grow only thanks to the fungi that nourish its roots.
- A plant thrives because bacteria fix nitrogen beneath it.
- The water that seeps downward carries minerals transformed by tiny microbes.

In some forests, mycorrhizal networks have been observed transferring resources between plants, though the extent and universality of these exchanges remain under study. Science moves carefully.

Interdependence is absolute.

But this fabric is fragile.

When we degrade soils, cut forests without restoring them, or seal the ground under concrete, the balance unravels: soil loses fertility and water, ecosystems collapse, and what once seemed solid begins to fall apart.

For millions of years, part of the organic matter created by life—ancient forests, microalgae, plants, and microbes—became trapped beneath sediments.

Over geological time, pressure and heat transformed it into dense, carbon-rich substances. There it remained, storing the Sun's radiant energy as chemical energy deep within The Earth.

Today, by extracting and burning it, we are returning that carbon to the atmosphere as CO_2—undoing in just two centuries what life took millions of years to balance, reigniting the same greenhouse effect that photosynthesis once helped to soften.

Soil, roots, and microbes are the true guardians of life.

To care for them is to care for ourselves.

*"All things are connected, like the blood that unites us.
We do not weave the web of life; we are merely a strand within it.
Whatever we do to the web, we do to ourselves."*

— Chief Seattle, Suquamish leader

Chapter 10

The Unity Of Life

The previous chapters revealed the cycles of matter and the invisible web of soil. Now it's time to step back and see the whole picture: there are no isolated parts. Everything that lives belongs to a single planetary network.

For a long time, humans imagined ourselves at the top of a pyramid—as if we were the final goal of evolution. But reality tells a different story: there are no tops or bottoms.

Life is a circle in which every species plays an irreplaceable role.

- The oxygen we breathe comes largely from microscopic algae.
- Trees exist because invisible microorganisms live inside and around their roots.
- Animals depend on pollinating insects and on the bacteria that dwell in their guts.
- Earthworms and microbes keep the soil fertile so that our food can grow.

Roughly half of the oxygen we breathe is produced in the ocean, thanks to photosynthetic plankton.

Once again, the tiny sustains the immense.

Each organism is a thread, and together they form a fabric that begins to tear when even one disappears.

A nightingale and an earthworm hold more of the future than any human machine, for they are irreplaceable pieces in the balance of life. And we, too, are part of that balance—not above it, but as one more note in the planet's symphony.

The unity of life is the deep law revealed by evolution: there are no absolute hierarchies, no disposable species. That unity gives us a simple ethical compass: what is true is what preserves the integrity of the web of life; everything else is noise.

Life endures because every form—from the tiniest to the most complex—participates in the same network.

When that fabric is torn—by pollution, deforestation, or invasive species—diversity collapses.

And when those networks weaken, it's not only species that disappear: the planet's ability to regulate water, temperature, and the carbon flowing through the atmosphere begins to falter. Ecological breakdown becomes climate breakdown.

The atmosphere breathes memories.

Every leaf, every river, every body is living memory of what once was.

A global report warns that up to a million species could be at risk of extinction in the coming decades if we do not change course. The challenge is to recognize it—and act accordingly.

Defending life does not mean protecting something outside ourselves; it means caring for the web that holds us all.

And so arises the question that leads us to the next step: if everything is connected, what happens when a single species—our own—breaks the threads of the web?

PART IV

THE FUTURE IS IN OUR HANDS

"Life is the way matter found to know itself.
Evolution is the path it took to do so."

— Carl Sagan, Cosmos (1980)

Chapter 11

Listening To The Earth Again

The Earth has never stopped speaking to us.

The wind carries whispers; the forest murmurs in a thousand tones. But for a long time, we stopped listening.

We became deaf amid the roar of engines, screens, and schedules.

The Earth does not shout—it whispers.

But when we fail to listen in time, the whisper becomes a hurricane, a drought, a fire.

Today, The Earth is raising its voice—not to punish us, but to awaken us.

When a hurricane tears through,
when a long drought cracks the soil,
when a forest burns where green once stood—
it is not nature "taking revenge."

It is her clearest language saying: **remember that everything is connected.**

Listening is not poetry—it is method.

Smelling damp soil, measuring how water seeps into the ground, following the flight of migratory birds—life speaks in data and in songs.

> *Listening also means distinguishing data from opinion.*
> *The Earth speaks in measurements, cycles, and trends;*
> *learning to read them is an act of responsibility.*

To listen to The Earth again is to open our senses to what has always been there:

- To notice how a tree blooms just when the bees need nectar.

- To see that soils rich in life produce healthier crops and cleaner water.
- To recognize that mangroves guard our coasts like living walls.
- To understand that every species plays a part in the symphony of life.

Listening is accepting limits.

If the river runs muddy, the watershed is asking for rest. If drought lingers, The Earth is reminding us that we have broken the water cycle.

And when the heat persists beyond what is natural, it is the carbon cycle warning that it too has been disturbed. If an insect disappears, it signals that a link in the chain of life is missing.

This is not about fear.

It is about **love.**

Love is the most powerful force we have,

because it moves us to care for what we value, to protect what we love.

And Earth is our only home—the place that gave us life and still offers everything we need to keep living.

To relearn how to listen is the first act of repair; without listening, there can be no covenant.

To listen to The Earth again is to accept the privilege of being alive in this moment of history.

Few generations have held a role as decisive as ours: we stand at the point where we can lose much... or awaken and regenerate.

Hope is not naïve.
Hope is action.
And action is born from truth:
when we attune ear and mind, we stop debating stories and begin responding to what reality asks of us.

Every seed planted, every river protected, every child taught to look at the sky in wonder is an act of love that changes the course ahead.

The Earth is speaking.

Do we want to listen?

"The Earth has music for those who listen."

— George Santayana, Spanish-American philosopher and poet

Chapter 12

The Human Dominion

For millions of years we walked like any other animal—gathering fruits, following rivers and migrations. We were part of the living layer, sustained by it, not apart from it.

Everything changed when we discovered agriculture. Forests were cleared to open fields. Rivers were diverted to irrigate crops. Wild animals were replaced by a few domesticated species.

With that came abundance—cities, art, science, and culture. But that expansion was always built upon the same invisible fabric of air, water, soil, and living beings that existed long before us.

The human presence on Earth is a paradox.

We can be moved by the beauty of a sunrise—and at the same time destroy the forest that receives its light. We can marvel at the song of a bird—and still invent machines that silence its home.

And unlike any other creature, we invented violence without hunger: killing for sport, for pride, for the pleasure of domination. Where others hunt to live, we have killed to prove power.

Every tree cut down is also a lost shadow, a weakened spring, soil that warms more quickly. Deforestation doesn't just erase landscapes— it alters rainfall, intensifies heat, disrupts cycles, and wounds biodiversity.

The same happened with accumulation.

No other species hoards more than it needs, because doing so would condemn its neighbor. We, instead, turned the obsession with accumulation into a badge of honor—praising those who gather while others are left empty-handed.

The essential truth is this:
we have never stopped depending on the invisible balance that sustains the
layer of life.
No human decree can suspend that dependence.
To mistake power for ecological immunity is to fall into fiction—
nature responds with facts.

Every loaf of bread, every sip of water, every breath confirms it.

And when we shatter ecosystems, we also break the invisible barriers that once protected us. Viruses and bacteria that once lived far from us now come into contact with our bodies. To care for forests and jungles is not only to protect trees—it is to protect our health.

But when we forgot that belonging, the climate lost its rhythm. By digging up and burning the energy that life had stored over millions of years as carbon buried deep underground, we released into the atmosphere CO_2 that was no longer part of the active cycle. We reignited an ancient heat that life had once managed to tame— more trapped warmth, more storms, more endless droughts.

Water, once the memory of landscapes, became scarce—or overwhelming.

Soils were exhausted.

The symphony fell out of tune, and with it, the chorus of biodiversity faded into silence.

Cities began to feel the echo of that forgetfulness.

Concrete and asphalt hold heat like open furnaces.

Without trees or breathing soils, neighborhoods become burning islands where life grows fragile.

We thought that imposing order in one place was progress, but in doing so we disordered the entire climate. What we called human dominion—what looked like triumph—was in truth forgotten dependence.

And yet, not all is lost.

The Earth responds when we stop forcing her: a restored mangrove calms the waves, a covered soil holds water, a shaded neighborhood cools itself again.

When we recognize the limit, we begin to hear once more the music we belong to.

True dominion is not over The Earth,
but over ourselves—
over greed, violence, and the hunger to possess.
And over the temptation of convenient lies.
Life does not respond to excuses—only to real change.

Human dominion shows us the limit.
Life, patient as ever, reminds us that response is still possible.

Chapter 13

A Personal Awakening

True change begins in silence—inside each of us.

Not in governments, nor in stock markets.

It begins the moment a person, in the midst of their own life, opens their eyes and realizes:

I am part of The Earth.

For a long time, we were taught to think small—to believe that our lives were limited to our house, our street, our job. That what mattered were the bills to pay, the problems to solve, the routines to keep.

But awakening is lifting our gaze. It is understanding that every breath connects me to the forests and oceans that produce oxygen. That every bite on my plate depends on living soils, rain, insects, and roots.

That my well-being is woven together with the well-being of millions of beings I will never see—but who quietly sustain me.

Personal awakening is not about guilt.

It's not about carrying the weight of the world.

It's an act of clarity: recognizing that what I do matters, and that my choices echo far beyond my walls.

Awakening also means choosing to be well informed.
Seeking reliable sources, checking facts, and refusing to spread
misinformation
is a concrete way to care for life.

That awakening becomes real through simple choices:

not killing for pleasure, but protecting life;
not hoarding beyond need, but sharing what is left;
not consuming energy without measure,
but living within the gift the Sun renews for us every morning.
Life has always thrived on flowing energy—not on energy extracted
and hoarded from The Earth.

When I walk down the street and see a tree, I can remind myself that the same life beating in me beats within it. When I open the faucet and water runs, I can give thanks that an ancient cycle still flows—one I didn't create, but on which I depend.

When I choose what to buy, what to plant, what to teach my children, I am tracing paths that either nourish or destroy the living layer.

To awaken is to move from **I** to **we**.

It is to understand that my well-being cannot exist without the well-being of The Earth. That my family's health is bound to the health of the soil, the air, and the water.

And it also means recognizing that not all carry the same weight. Some peoples endure greater droughts, hurricanes, and pollution—though they were not the ones who broke the balance. Personal awakening deepens when we understand that caring for The Earth also means seeking justice for those who have waited too long.

Forgetting that life learned how to sustain itself is what puts it at risk.

Psychology confirms it: what moves the human being is not constant fear, but a sense of belonging and purpose.

When we feel that we are part of something larger, our lives gain clarity and strength.

That is the personal awakening: the certainty that my life has meaning because it is connected to the life of all others. I am not alone. None of us are.

We are part of a larger organism, an immense web inviting us to care for it.

And this is our privilege:

to live in the historical moment when we can still choose. Because the memory of balance that pulses within The Earth can also be reborn within us.

Just as soil regenerates when it is covered, our hearts regenerate when we care for them.

The symphony begins to be heard again.

We breathe deeply. The noise subsides.

And then, what has always been there appears—a gentle music not coming from outside, but from the fabric that holds us.

The soil pulses with millions of invisible voices;
water keeps time as it seeps;
leaves tune light into sugar; animals enter and exit with their own rhythm.

And we, at last, remember our part in the choir.

It is not a perfect symphony—it has silences, dissonances, and lessons. But when we listen, what matters falls into place: to care for what sustains life.

This is the personal awakening: to move from spectators to participants. And from here, we cross the threshold into shared action.

The music calls for agreements, hands, direction.

What follows is a covenant—not signed in ink, but in actions repeated until they become culture.

Chapter 14

Realistic Hope

Awakening is not enough if it doesn't turn into action.

An open heart needs hands that move.

And here emerges a word that often seems fragile, yet is the strongest of all: **hope**.

This is not naïve hope—the kind that looks away from destruction. It is realistic hope: the kind that faces the wound and still chooses to act.

Because we have seen it again and again: when we make room for life, life responds.

A devastated forest greens again when given the chance.

A polluted river begins to cleanse itself when we stop polluting it.

A species thought lost returns when it finds refuge.

Realistic hope is born from learning from our mistakes.

We discover that killing for pleasure destroys more than just the prey that limitless accumulation leaves our neighbors empty, that exploiting energy without restraint throws the planet's climate out of balance.

But we also discover that each opposite gesture holds immense power: to protect instead of harm, to share instead of hoard, to live with enough instead of devouring what we don't need.

Regeneration begins when we stop consuming borrowed energy from the past and learn to live with the energy nature offers us in the present.

Realistic hope looks straight at measurable results.

Its compass is the truth of processes, not the wishful thought that "maybe nothing will happen."

Science confirms it with concrete examples:

- Tropical forests once cleared now show young trees thanks to community efforts.
- Whales once near extinction sing again in the oceans.
- Polluted lakes that were factory sewers are now living waters, repopulated by fish and birds.

This is not evasion—it is planning.

It means recognizing that there is no time to lose,
but also that every action counts.

It is not about waiting for miracles—
the miracle is already here:
the capacity of life to regenerate when we care for it.

Fear paralyzes us; hope moves us.
Fear says, "It's too late."
Hope answers, "Every day counts."

Because when we plant, restore, and protect,
the world responds with astonishing speed:
the soil regains its scent of forest,
bees find flowers where there was concrete,
children learn to look at the sky and see a possible future.

Realistic hope reminds us that regeneration is not only technique—it is also justice.
Those who have polluted the least are often the ones who suffer the most—from droughts, hurricanes, or lost harvests.

True hope cannot be a privilege for a few;
it must take root in every neighborhood,
in every field,
in every community.

Realistic hope is not a fragile illusion—it is a proven certainty.
The certainty that when we act together, life responds.

At this point, the symphony grows stronger again.
Personal awakening becomes shared hope.
The intimate becomes collective.

This is the step that prepares us for what comes next:
not just to hope, but to commit.
Not just to dream, but to make a pact.

Because realistic hope reminds us that regeneration is possible,
and the covenant with The Earth will show us how to sustain it.

"Hope is a verb with its sleeves rolled up."

David W. Orr, Ecologist and Professor of Environmental Studies

Chapter 15

Two Paths Before Us

In its rush to fix the disorder it created, humanity has begun to imagine giant machines that could capture and clean the air, artificial clouds to block the sun, fertilizers thrown into the oceans to absorb more carbon.

They call it geoengineering—an attempt to correct the course of The Earth with the same mindset that broke its balance.

But The Earth is not a machine.

It is a living fabric.

And when you tug on a single thread without understanding how it is woven into all the others, you risk unraveling the whole tapestry.

Geoengineering promises speed, but it is a dangerous illusion—moving pieces without understanding the entire game. We have tried that logic before: believing we can borrow energy from the past without consequence, instead of learning to live within the limits of the present.

The clearest example lies in the Amazon.

That immense forest—keeper of rains and carbon—is nearing a point of no return. If too much of its cover is lost, it could turn into dry savanna. That would not just be a change of scenery: millions of species would vanish, and the climate of an entire continent would be altered. This is the risk of thinking we can manipulate systems we do not fully understand.

History warns us: when we kill for pleasure, when we hoard without limit, when we exploit energy beyond what is needed, we always unleash a greater disorder than the one we set out to solve. The same logic of excess cannot be the one that saves us from the crisis.

The other path is humbler, slower—and **truer**: to learn from The Earth itself.

She carries in her memory millennia of adjustments—storms,
droughts, ice ages, and renewals.
She has always known how to sustain life.

That knowledge isn't written in a lab or an algorithm.
It is written in the sap of trees, in the cycles of water,
in the dance of soil microbes,
in the breathing of forests and oceans.

We call this path **nature-based solutions**:

- Restoring forests that regulate rainfall.

- Healing soils so they can store carbon and water.

- Protecting mangroves that shield coasts and nurture marine life.

- Weaving corridors of biodiversity so species are never alone.

- Farming with regenerative methods that imitate the logic of the
  forest instead of forcing it with synthetic chemicals.

This does not mean turning away from all technology.
It means recognizing that the foundation of any lasting solution is to
accompany nature's processes, not replace them.
Science has much to offer—but only when it learns first from the
ancient wisdom of The Earth.

One path promises shortcuts, even if it denies reality.
The other follows the truth of the cycle, even if it takes more time.

The difference is clear:
one seeks to impose,
the other to collaborate.

Those of us who created the problem by breaking nature's harmony
will not solve it by repeating the same arrogance.

The way forward is to help The Earth in her own processes,
not to substitute them with artifices.

Life has spent billions of years learning how to stay alive.
Only life knows how to sustain life.

The real future cannot be built on smoke and illusions.
It must be built on roots, seeds, clean water, and fertile soil.

Science already confirms it:
when we return life to The Earth, she answers with generosity.

And poetry reminds us:
when the forest breathes, we all breathe.

Today, two paths lie before us:

one of artifice, promising power;
and one of humility, offering hope.

The first is short, fragile, and full of risks.
The second is long, patient, and fertile.

One seeks to dominate.
The other learns to cooperate.

And though the temptation of speed will always exist,
experience teaches us that the safest path
is the one grounded in The Earth herself.

Only then can we choose a future that sustains life.

*"Something is right when it tends to preserve the integrity, stability, and beauty of
the biotic community. It is wrong when it tends otherwise."*

— Aldo Leopold, ecologist and land philosopher

Chapter 16

The Threshold

There are moments in history unlike any others.
They do not feel like the past, nor are they yet understood as the future.
They are thresholds—
doors that open only once.
And we stand now before one of them.

For billions of years, life was learning—
not with words, but through processes—
how to sustain a living planet.
Every organism that ever existed, every cycle that closed, every molecule taken and returned,
was part of a silent conversation between The Earth and life itself.

And for almost all that time, no one was watching.
No one could know.

Life simply was.

Life simply continued.

But we can see.
We are the first species capable of looking back
and understanding what made it possible for us to be here.
And that awareness brings with it a responsibility we cannot ignore.

Until very recently, humanity walked without understanding the ground beneath its feet.
We lived, worked, celebrated, suffered, were born and died upon an Earth that seemed infinite and stable.
We took for granted the air, the water, the soil's fertility, the forests, the seas.
We did not know—none of us knew—
that all of it was the result of delicate cycles that took eons to stabilize.

We arrived late to understanding.
But we arrived.

Today we stand before a threshold unlike any before:
the moment when we know, without doubt,
that we can alter the very pillars that sustain life.
And it is no longer possible to look away.
No longer possible to live as if we did not know.

The world is speaking to us—
through rising seas, record-breaking heat,
extreme droughts, exhausted soils, vanishing species, dying corals—
and its language is clear:
we have touched the systems that keep everything alive.

This threshold is not tragedy.
It is not punishment.
It is truth.

And **truth** here is not an idea;
it is data, causes, consequences—
what climate, water, and soils are already saying.
A truth that awakens us and calls us to choose.

We can go on as if nothing had changed, ignoring what we know.
Or we can cross with awareness, humility, and courage—
recognizing that our intelligence was never meant to dominate life,
but to protect it.

We are the first generation to truly understand what is at stake—
and perhaps the last with the time and capacity to act.

That is not cause for fear.
It is cause for dignity.

We enter this new stage not because we wish to,
but because reality has brought us to the door.

What we do now will be remembered by those who come after—if
they come—
as the moment when humanity looked in the mirror and decided who
it wanted to be.

Choosing who we are begins with refusing to lie to ourselves about what is happening.
From there, every covenant makes sense.

This threshold does not ask for perfection.
It asks for awareness.
It asks for honesty.
It asks for will.

Breathe.
Look at the world around you.
Listen for your place within it.

Perhaps remembering who we are is not about thinking,
but about letting life speak through us.

This is the moment.

The door is open.

Let us cross it together.

$6\,CO_2 + 6\,H_2O \rightarrow C_6H_{12}O_6 + 6\,O_2$

$2\,C_6H_{14} + 19\,O_2 \rightarrow 12\,CO_2 + 14\,H_2O$

$C_6H_{12}O_6 \rightarrow CH_4 + C(s) + H_2O + C_nH_m$

Chapter 17

When We Opened The Vault Of The Past

For billions of years, Earth worked without rest.

It didn't think or desire: it followed the laws of life and the universe. Everything that breathed, grew, or died was part of the same cycle: daylight, air, water, and soil, and then night; that simple pulse held the complex together.

At first, the air was loaded with carbon dioxide and methane. It was a planet too hot for creatures like us. Then came the great turn: photosynthesis. It didn't "create" energy; it transformed sunlight into chemical energy stored in bonds. Every sugar made by a leaf or a photosynthetic cell was sunlight saved as food. With every molecule, a bit of carbon left the air and entered the web of life.

When those organisms died, part of their matter sank to the seafloor or was buried by sediments. Over time, pressure, and heat, that material compressed into coal, oil, and gas: ancient sun locked in the bonds of hydrocarbons. No one planned it. It wasn't a chest prepared for the future; it was a blind consequence that, by removing carbon from the air, cooled the common home. As it cooled, complex life—the life that breathes, thinks, and dreams—found room to exist.

That was the turning point.
Forgetting that story—or distorting it—does not change
the chemistry of the air.
Remembering it is honoring the truth that allowed us to exist.

For eons, that balance held.

Until we opened the vault of the past.

We pulled coal from its seams, oil from its rocks, gas from its deep cracks, and in seconds returned to the sky what took geological ages to store. Every barrel of oil is light from millions of years ago turned back into today's heat. Every combustion releases CO_2 that was outside the atmospheric cycle. By returning it, we recreate ancient climates—hotter, more unstable, more violent for everyday life.

Numbers vary across studies, but the direction doesn't change: more carbon in the air → more energy trapped → more heat → more imbalance.

Oceans warm and acidify. Corals—the forests of the sea—turn white. Soils lose sponge-like structure and moisture. Rivers lose their meanders and rush past without talking to their banks. Storms turn up the volume. Glaciers retreat; the poles melt. Our lungs breathe air with particles and gases the body can't handle for long. Earth answers with its laws: it doesn't punish; it responds.

This isn't ideology; it's physics.

Understanding it isn't about guilt; it's about clarity. Because now we know—and knowing obligates us.

Life's Accounting: today's paycheck vs. yesterday's savings

The biosphere keeps simple, strict books. Each day brings today's paycheck: sunlight that arrives, wind that moves, rain that falls, photosynthesis that captures. With that paycheck we pay for everything: growth, repair, reproduction, pollen's trip, birdsong, human work.

Yesterday's savings are something else: energy from fossils formed over millions of years. When we burn them, we spend savings that aren't replenished on the timescale of life. That sudden injection speeds everything up. And by speeding up, we knock things off balance.

Let me say it again, another way:

- Today's paycheck is today's sun, captured by leaves, winds, and currents.
- Yesterday's savings are yesterday's sun, stored underground.
- Living on the paycheck keeps the cycles in rhythm.
- Living off the savings breaks the rhythm: we do more than life can put back.

A third way, in a home image: today's paycheck is the bread baked each morning; yesterday's savings are a pantry full of thousand-year-old loaves. Eating from the pantry doesn't just empty old bread; it changes the house's rhythm: cooking gets rushed, meals multiply, portions grow, the oven is forgotten. The family thinks it "produces more," but in truth it lives faster than it can sustain.

Acceleration: when all of life hits the gas

With dense, cheap energy we multiply speed, volume, and distance. We extract more materials, move things farther, consume faster. Even if technology improves and an engine gets more efficient, the rebound effect shows up: because each use is cheaper, we use more—and total use rises.

We fix one leak and open three more faucets.

Said in farm language: we buy an "efficient" pump and, since it uses less per hour, we leave it on twice as long. Total consumption doesn't go down; sometimes it goes up. That's rebound. Multiplied by millions of choices, it becomes more combustion, more CO_2, more heat.

Acceleration is felt in the body: everything "has to be now."

It's seen in the landscape: harvesting too early, stripping ground cover, straightening rivers so water hurries. We live as if soil were a hard court, not a living sponge.

The rhythm of day and night

For most of human history, we worked with light and rested with dark. Opening the vault changed that choreography: electricity and fuels stretched the day. Shift work that never ends, screens that never sleep, cities lit at midnight.

Our bodies, however, are ancient. Every cell carries a circadian clock. When that clock doesn't match the sky—night shifts, bright blue light late, meals at odd hours—the body protests: broken sleep, irritability, hormonal imbalances, illnesses that build quietly. What we've done to the climate we repeat in ourselves: out of rhythm outside, out of rhythm inside.

Once more, clearly:

- Day is for capturing energy; night is for saving and repair.
- Burning savings turns night into day and drains the body's reserves.
- If we resync with light—no fanaticism, just sense—we gain health and use less.

And a third time, with a simple gesture: turn off one unnecessary light tonight. Feel how the house slows down. Notice how silence gathers strength.

Limits that protect

Limits aren't punishment; they're shape. The bank gives the river shape; skin gives the body shape. Without a bank, the river spreads; without skin, the body is hurt. Limits on consumption, speed, and scale protect what we love. They don't demand perfection; they ask for consistency. The pact lives in what we repeat.

So we repeat what matters, with small variations, so everyone can find it:

- First way: we live on today's paycheck → we spend less, we regenerate more.
- Second way: we close the vault gradually → we stop spending ancient savings.
- Third way: we slow down → we give soils and waters time to heal.

Soil, water, and community: the living infrastructure of energy

The transition isn't just panels or turbines; it's landscape and culture.

Soil. Palm flat on the ground. Cover it, feed it, keep it spongy and forest-scented. Covered soil stores carbon, holds water, and cools the place. Where there was dust there will be mulch; where there was hurry, patience.

Water. Invite it to stay. Harvest it when it falls, let it infiltrate where it runs, give curves back to what we straightened. Rivers need meanders to slow their thinking. When water has time, life has soil.

Community. What a family does is a seed; what a neighborhood agrees on is a garden; what a city sustains becomes a landscape. We learn by watching, teach by doing, and celebrate by sharing. The energy we don't spend by being near is the cleanest of all.

Energy justice

A few opened the vault; the heat is shared by all. Those who burned the least often suffer the most. Repair requires fairness: relieve first the communities already carrying pollution, extreme heat, and precarious work. Repeated truth: a transition without justice isn't a transition; it's a machine swap.

How do we close the vault without dimming life?

It's not about going back to caves; it's about returning to the rhythm.
- Sufficiency before efficiency. Cut needs, right-size scale and speed; then improve technology.
- Live with daylight. Work and school in daytime when possible; minimal, warm night lighting; rest as part of the pact.
- Relocalize the essentials. Food and materials close to home: less transport, more community.
- Redesign landscapes. Covered soils, wildlife corridors, rain harvesting, meanders restored.
- Renewable electricity for what's essential, and lower demand so it truly covers it.
- Every purchase is a vote. Avoid products that travel with deforestation, chemicals, and waste; choose local, regenerative, and fair.
- Consistency. Habits that turn into culture. Don't demand perfection; repeat what works.

Again, a pocket version to memorize: **less, closer, slower, more alive**.

Teaching the pact

A child walks with her grandmother. They gather dry leaves and spread them over the planter. Without knowing the word "mulch," she understands the soil shouldn't be naked.

At school, a metal cylinder set into the ground marks the rhythm of a simple experiment: a drop falls and sinks; the stopwatch runs; wonder does too. That cylinder teaches the logic of today's paycheck: if water infiltrates well today, there will be water tomorrow.

In the plaza, two neighbors unroll a hose, plant a tree, and—without meaning to—plant a conversation. The tree will do its part; the conversation will build culture.

A short ritual to remember the way: set a container under the eave and catch the first rain; plant one seed in the ground and another in someone's heart; tell what you saw when you looked slowly at the world that holds you. Tomorrow, repeat.

Choices that tune the rhythm

- •Turning off unnecessary lights at night isn't symbolic: it re-syncs body and neighborhood.
- • Covering one square yard of soil today lowers tomorrow's fever.
- • Giving one curve back to a channel returns memory to water.
- • Choosing what's nearby shortens energy distances.
- • Accepting limits protects the shape of what we love.

And in case a fourth repetition of the core idea helps: life knows how to live with the sun that arrives each day. When we spend what belongs to yesterday, we speed up beyond what the cycles can restore. Closing the vault isn't going backward; it's respecting the process that brought us here.

We're not late if we slow down. We're not late if we lay our palm on the ground again and listen to its double bass. The harp of water keeps tuning meanders; the violin of light returns every morning.

Our task isn't to finish a perfect work: it's to learn our part and play on time, again and again, until the planet's music breathes without hurry.

Because we know what's at stake—with scientific certainty and human clarity—we can no longer pretend not to know.

We close the vault.

We live with the sun of today.

And we turn that **decision** into **habit**, and that habit into **culture**.

"What has become clear from the science is that we cannot burn all of the fossil fuels without creating a very different planet."

James Hansen, Climate scientist and former director of NASA's Goddard Institute for Space Studies

Chapter 18

Returning To *Us*

The time has come to return.
To look at one another again.
To recognize ourselves.
To remember that we have never been alone.

For a long time, we believed the wrong stories—
that strength lies in independence,
that success is personal,
that human beings rise, think, create, and survive alone.

But that is not the story of life.
It never has been.

From the first cell that shared nutrients,
to the first forest that held the humidity of a watershed,
life has always thrived in community.

Cooperation was not a philosophical idea—
it was biology's winning strategy.

Plants speak to each other underground.
Bacteria organize themselves to survive in colonies.
Bees live for the hive, not for themselves.

And we, though we may have forgotten,
are born defenseless,
and grow upheld by hands, voices, and loving eyes that teach us the
world.

Life has always been a we.

But one day we disconnected.
We built cities of concrete and metal,
and planted ourselves in pots—
separated from one another,

far from The Earth, from the natural rhythm,
from the living web that sustains us.

Because a plant can survive in a pot.
It can be watered, it can grow, it can even bloom for a while.
But it does not belong.

In a pot, its roots cannot travel or meet others;
they do not share nutrients or chemical signals;
they do not participate in the underground conversation that sustains
a forest.

A potted plant lives—
but it does not thrive as part of life.

That is what happened to us.
We planted ourselves in invisible pots:
isolated homes, disconnected routines, lives locked inside the idea of
"me"—and we forgot the web that holds us.

Life in a pot is a reduced life.
It endures... but it does not expand.

And now, the time has come to replant ourselves in the living soil of
"us."

Returning to common ground also means returning to a common
language:
shared evidence, genuine listening, and agreements rooted in reality.

In our isolation, we lost the memory of our origin.
We began to believe that artificial meant progress,
that fast meant good,
that more meant better.

Until life itself reminded us of the truth:
we are not machines—
we are organisms.

And organisms do not live alone: they breathe in networks.

Returning to us does not mean renouncing who we are.
It means remembering that we are more when we are with others.

Look at your own body:
no organ lives on its own.
The brain does not feed itself.
The heart does not make its own oxygen.
Everything works because everything cooperates.

The Earth works the same way.
Oceans, forests, soils, currents, winds, pollinators, fungi—each does
its part.

No more, no less.
And because of that, life continues.

The inheritance of life is not kept in vaults.
It lives in ecosystems.

This is not a time for solitary heroes.
It is a time for connection—
for hands that plant,
for minds that think,
for youth that imagine,
for elders who remember,
for scientists, farmers, artists, engineers, teachers, fishers, mothers,
workers, cooks, doctors—
for the girls and boys who still know how to listen to The Earth.

Each one does their part, from where they are.
Not what belongs to someone else.
Not the impossible.
Their part.

That is what it means to return to us.

Because we don't all have the same tools—
but we all have responsibility,
and we all have power.

Returning to us does not demand perfection.
It calls for humility.
It calls for belonging.
It calls for recognizing that we are part of life,
and that life sustains us when we learn to sustain it in return.

We did not come to impose.
We came to learn again.
To listen to the rhythms we have forgotten.
To feel the earth beneath our feet.
To remember that air is a shared gift.
To honor water as if our breath depended on it—because it does.

To return to us is to look at another person—and at the planet—and
say:

"Your well-being is part of mine."

Because it is.

When a community grows food together, there is nourishment.
When we restore a river, there are fish—and there is life.
When we defend the soil, there is future.
When we share knowledge, there is hope.
When we care for a child, we care for the whole world.

To choose life is to choose us.
To choose us is to come home.
To return to us.

And when we do, The Earth can breathe again.

Because The Earth does not ask for sacrifice—
it asks for coherence with life.

No one can save the world alone.
But each of us can hold a single thread—
and when those threads come together,
they form a web.

And the web carries what none of us could bear alone.

This is the moment.
Here is where we stop being spectators.
Here is where we become community again.
Here is where we meet one another once more as a species—
and lift our gaze, together,
toward the life that sustains us.

To return to us is to return to life.
And we are ready.

*"When we try to pick out anything by itself,
we find it hitched to everything else in the universe."*

— John Muir, Scottish-American writer, naturalist, and conservationist

Chapter 19

The Covenant With Life — Addressed To You

You've come this far.
And that means something.

Not everyone does.
Not everyone stops to look at the world with open eyes,
to feel both the weight and the beauty of being alive,
to understand the gift and the responsibility
of existing on this one-of-a-kind planet.

But you did.

And now that you know,
it's your turn to decide.

This is not a covenant with me,
nor with any group, institution, or flag.
This covenant is between you and life itself.

There are no intermediaries.
No requirements.
No registration.
No applause, no recognition, no membership to earn.

There is only one simple, profound question:
Will you walk with life, or against it?

I don't ask you that as a judge, or as a teacher,
or as someone with authority.
I ask you as another living being who woke up in time—
who realized what truly matters,
and who understood that life—all life—
is the most sacred thing in the universe.

Because if there is a sanctuary, it is this planet.
If there is a miracle, it is being alive here, now.
And if there is a duty,
it is to care for the conditions that make life possible.

Nothing more.
Nothing less.

The Earth does not ask for perfection.
She asks for presence.
She asks for respect.
She asks for coherence.

And in this moment, she asks for action.

She asks us not to destroy what sustains us.
She asks us to honor what gave us the chance to exist.
She asks us to walk with her, not against her.
She asks us to care for the ordinary things we take for granted—
the very things that make life possible.

And now you know the truth:
Only life can care for life.
And only truly intelligent life can protect life.

Intelligent life is not the one that accumulates data—
it is the one that recognizes truth.

Because nature cannot be deceived by lies:
if something harms her, it will continue to harm her
no matter how many say otherwise.

Lies can fool humans—
but they cannot fool The Earth.

Natural laws do not obey decrees or propaganda.
They obey only **truth**.

That is why those who choose to protect life
must also protect truth.
Not because it is an abstract virtue,
but because without truth
no form of life can endure.

And in this time,
when misinformation is used as a weapon,
defending truth has become a form of love.

Let us never forget:
The Earth does not listen to speeches.
She responds to actions.
And actions are the clearest voice of truth.

Life favors those who favor life.
Life expels those who destroy it.
That is not vengeance—it is natural law.

Today you hold a clarity
that generations before you did not have.
You are part of the first humanity
to understand the cycles that sustain existence—
and perhaps the last with the power to protect them.

That is not a burden.
It is a historic privilege.

And today, here—
without witnesses or rituals,
without audience or noise—
you can make the most important covenant of your life:

To choose to stand on the side of life.

From wherever you are.
With whatever resources you have.
At your own pace.
In your own reality.
Without comparison.
Without asking for permission.

Caring for the soil beneath your feet.
Caring for the water that sustains you.
Caring for the life that surrounds you.
Caring for your own humanity, your meaning, your dignity.

Without fear.
Without shame.
Without waiting for someone else to go first.

Because you are already awake.
Because you already understand.
Because you already know.

And now, with a clear heart and open eyes,
you can say to yourself:

I choose life.

I walk with life.

I defend life.

Not to be a hero.

Not to be a martyr.

But because that is what it truly means to be human.

And if ever you doubt, remember what matters most:
Everything that grows, breathes, feels, beats, and blooms
is part of you.

And you are part of it.

This is your place in the universe.
This is your time.
This is your calling.

Welcome.

You are already on the side of life.

Now, let's walk together.

"I want you to understand that we are part of the natural world. And even today, when the planet is dark, there still is hope."

Jane Goodall, Final interview, Famous Last Words (Netflix, 2025)

Chapter 20

The Continuity We Are

You've reached the end of this journey.
And if something inside you has been kindled—
a feeling of certainty, of return, of tenderness, or of urgency—
it was no accident.
It was your body responding.

Because long before we understood through words,
we already carried within us a living wisdom—
not learned, not heard, not observed,
but silently transmitted through the continuous current of life.

Life forms are not invented anew with each generation.
Life is not created over and over again.
Life is passed on:
from cell to cell,
from body to body,
from species to species—
through millions of years of uninterrupted continuity.

What has kept us alive—
the impulse to breathe, to seek shelter, to reproduce, to cooperate, to
protect life—
is not something we are taught.
It is coded within us.
It is part of the great flow of existence that brought us here.

Life remembers what the mind forgets.
And within that silent memory, we remain continuity.

Birds don't learn how to build nests;
they build them because the pattern is imprinted in their bodies.
Plants don't reason that they need pollinators;
they open their flowers because the life within them knows
that other beings will come and be nourished.
Bees don't make honey merely to produce food;

in the hive is imprinted the act of storing sustenance,
of reproduction,
of ensuring the persistence of a being
that sustains us as part of the ecosystem.
Roots don't calculate the presence of fungi;
they cooperate because that ancient bond
has supported both for millions of years.

We, too, come from that web.
And when we act in harmony with it, we feel peace, belonging,
purpose.
But when our actions drift away from what protects and sustains life,
a deep unease awakens within us.
We feel alone, lost, disconnected—
because deep down, we know we've stepped off the path that brought
us here.

Even though science is only beginning to explore this continuity—
through studies on biological inheritance, epigenetics,
and the transmission of unlearned functional patterns—
life already knew.
And so do you.
Not because you remember it.
But because you are it.

So if something in these pages made sense to you
without you knowing exactly why—
if you felt, "This is true,"
even though no one ever taught it to you—
perhaps it is because it already lives within you,
like breath,
like heartbeat,
like the instinct to care for what sustains you.

I invite you to read this book again from that living awareness—
not from the mind that reasons,
but from the body that pulses.

Because to understand The Earth
is also to recognize that we are part of that unbroken current of life.

That is what is **sacred**.

And that here and now,
we hold both the possibility—and the responsibility—
to decide whether we will continue to be part of it...
or separate ourselves from its pulse.

If this book touched you, you are not alone.
There are many of us, in different corners of the world,
also remembering with our bodies and hearts.

We have created a living network—protected, intimate, and committed—
to stay connected:
a space to share ideas, actions, projects, and questions,
a place to care for one another and for The Earth.

If you feel the call,
you can join the group described on the last page of this book.

This is not the end.
This is only the beginning—a beginning that, for us, must be filled with **hope**.

And Now That You've Finished...

Open your eyes.
Look around you.
Feel the ground beneath your feet.
Your body is still here—alive, aware, capable of caring.

The story you've just read doesn't end with the last page.
It is a seed. A calling.
A memory awakening.

Remember: you are not alone.
There are many of us who are remembering too—
who feel, in our bodies, the same **truth** you have felt.
And we know that life can only be cared for **in connection.**

The Earth cannot care for herself.
Nor is it enough to protect a single species.
Every action you take—no matter how small—
touches the entire fabric of life.
Because everything that breathes, cooperates.

Breathe again.
Connect.
And take a step, even a small one.

This book is not an ending—
it is the root of a new beginning.

Because to care for The Earth
is to care for the life that also lives within you.

Letter To Those Who Believe In Creation

Dear friends,

I write to you as an equal—
a neighbor, a sister, a fellow traveler in this same living blue home.
I do not come to debate; I come to extend a hand.

Science shows us how the world works;
faith reminds us why it matters.
If we believe that life is the work of a Creator,
then the conclusion is clear: to care for Creation is an act of faith.

The first calling of any church is not to condemn,
but to protect life.
The ancient mandate to "keep and cultivate" remains alive:
to defend rivers, soils, mangroves, forests, coasts, and skies.

At times, without meaning to, we speak of the greatness of the
Creator's work
but leave our hands idle.
Today, the planet needs fewer speeches
and more hands that repair and celebrate life.

Some wait for a miracle to solve everything.
But God has already given us the broom—
intelligence, hands, and hearts.
True faith is shown by using them to care.
The miracle is that we still have the chance to act.

No generation before ours has had the knowledge we now possess.
And with knowledge comes responsibility.
Whoever understands and is able to act, must do so—
even through small but steady steps.

I offer a few concrete and replicable gestures:

- Reduce the energy use of temples and install solar panels
  wherever possible.

- Turn church grounds into gardens of biodiversity and shade.

- Organize clean-up and reforestation brigades along rivers and
  coasts.

- Create solidarity funds for energy efficiency and rainwater
  harvesting in vulnerable homes.

- Dedicate one month each year to a "Season of Creation":
  sermons, composting workshops, gardens, and reuse projects.

- Prepare community resilience plans for hurricanes and droughts.

- Support regenerative food production and declare temples
  "climate refuges,"
  because every community can become a seed of restoration.

None of this asks us to leave faith at the door;
on the contrary, it means to embody it.
Every drop of clean water is a prayer.
Every tree planted, a psalm.
Every living soil, an amen spoken with our hands.

Love proves itself through care.
Let us make each temple a beacon,
and each act of worship a promise fulfilled—
in the garden, by the riverbank, in the community kitchen.

Let children see how faith becomes garden, shade, clean water, refuge,
and bread.

There is no time to lose,
but there is still time to begin—
here, now, together.

In this way we sign, with our actions,
the covenant that both faith and The Earth call us to fulfill.

With respect and hope,

A fellow traveler on this shared home we call Earth

Letter to the Scientific and Educational Community

Dear colleagues, educators, and scientists,

We have devoted our lives to unraveling the mysteries of the universe—to celebrating the human capacity for discovery and technological wonder. But we must also have the humility to acknowledge our limits.

We live on a planet that is a living, interwoven system—connected in ways we are only beginning to understand.

The fragmentation of our disciplines has produced partial solutions, sometimes with serious unintended consequences. No generation has ever possessed so much validated knowledge; that privilege is also an ethical burden. If we clearly understand what is happening, we cannot remain silent, nor can we stand still.

The most noble task of science is not to impose an artificial balance, but to learn from the balances The Earth has achieved through her own processes. That requires breaking barriers between disciplines, avoiding sterile rivalries, and keeping at the center what truly matters: the preservation of life.

Our duty is not confined to laboratories or classrooms. Every educated person can be an ally in restoring balance. The solution will not come from isolated discoveries, but from a collective transformation of mind and culture.

I invite you to use data and evidence not only to publish papers, but to inspire communities, guide policy, and sustain concrete action. Let us teach with breadth, humility, and commitment. Let us remember that knowledge becomes wisdom only when it is translated into care.

Because the true covenant with The Earth
is not signed with words,
but with actions that protect life.

With respect and hope,

A fellow traveler on this shared journey of learning

What Truly Matters

To those who today sustain destructive industries,
or who, from positions of power, choose to ignore the cry of The
Earth—
history will not remember their speeches or their excuses.
It will remember their actions, and their silences.
The consequences will reach their own lives, and those of their
descendants.

But I do not write these words for them.
I write for those who still wish to listen,
for those seeking coherence,
for those who long to protect life and sow the future.

In the end, what is at stake
are not our passing differences
nor the arguments that divide us.
It will not matter which party won an election,
which ideology prevailed,
or who was praised or condemned from a pulpit.
If we keep breaking the common ground—
the soil, the water, the air, the living fabric that sustains us—
the damage will reach us all.

The only thing that truly matters—
and the only thing history will remember—
is whether we were able to **keep life alive.**

GLOSSARY

Anthropocene: A term some scientists use to describe the current age of The Earth—when human activity has altered the planet so deeply that it has left lasting marks in the atmosphere, oceans, and soils, as if we were a geological force.

Archaea: Tiny microorganisms similar to bacteria, but with unique traits. They live in extreme environments such as hot springs or salt lakes, showing that life can adapt to almost any condition.

Atmosphere: The layer of gases that surrounds The Earth. It contains oxygen, nitrogen, and other gases that make life possible by regulating temperature, filtering solar radiation, and keeping water in its liquid state.

Atom: The smallest particle of a chemical element that retains its properties. It is made of protons, neutrons, and electrons, and forms the molecules of all matter.

Biodiversity (biological diversity): The variety of living forms in a given place—from microbes to trees, insects, and large animals. The greater the biodiversity, the more stable and resilient an ecosystem becomes.

Biotic Community: A group of living organisms that inhabit and interact in the same place. Together with their physical environment, they form an ecosystem.

Big Bang: The name of the scientific theory that describes the origin of the universe. It refers to an initial moment, about 13.8 billion years ago, when all matter and energy were concentrated and began to expand, giving rise to space and time.

Biosphere / Layer of Life: The zone of the planet where life exists. It includes the lower part of the atmosphere, the land surface, bodies of water, and fertile soils. It is a thin "living skin" where all life takes place.

Carbon Cycle: The process by which carbon circulates among the atmosphere, oceans, soil, plants, and living beings. It keeps the climate in balance and allows the formation of organic matter.

Carbon Dioxide (CO_2): A gas that is naturally part of the atmosphere, but in excess causes global warming. It is produced by living beings during respiration and by the burning of fossil fuels.

Cambrian (Cambrian Explosion): A period in Earth's history that began around 541 million years ago. During the Cambrian, many forms of life appeared suddenly with new "body plans," leaving a diverse fossil record in the seas.

Chemical Energy: Energy stored in the bonds of molecules. It is released through processes such as respiration and combustion.

Chloroplasts: Organelles inside the cells of plants and algae where photosynthesis takes place. There, sunlight is transformed into chemical energy that feeds the plant.

Composting: A natural process in which microorganisms break down organic matter (leaves, food scraps, branches) into a fertile material called compost or humus.

Cosmos / Universe: Everything that exists—matter, energy, space, and time. It includes everything from the smallest particles to the most distant galaxies.

Cyanobacteria: Tiny microorganisms similar to bacteria that perform photosynthesis. They were among the first living beings to release oxygen into the atmosphere, transforming it and making life as we know it possible.

Deforestation: The loss of forests through logging, fires, or agricultural expansion. It destroys the homes of thousands of species, reduces the oxygen we breathe, and accelerates climate change.

Decomposition: The process by which dead organisms and organic waste are broken down into simpler matter, returning nutrients to the soil and closing natural cycles.

DNA / RNA: DNA is the molecule that stores the information of all living beings, written in a code of four chemical letters. RNA is a similar molecule that helps read and use that information. Together, they allow each cell to build proteins and sustain life.

Ecosystem: A community of organisms that interact with one another and with their physical environment (soil, water, air). Each ecosystem maintains a dynamic balance that sustains life.

Ediacaran: A period in Earth's history, just before the Cambrian, when soft-bodied organisms appeared and left imprints in rocks.

Endosymbiosis: An intimate relationship in which one organism lives inside another to mutual benefit. It explains the origin of mitochondria and chloroplasts.

Energy: The capacity to produce change or movement. In nature, it appears as sunlight, heat, electricity, the movement of water, and the chemical energy stored in living organisms.

Entropy: The natural tendency of energy and matter to disperse and move toward disorder. In the universe, entropy always increases, even though ordered structures can form in certain places.

Evolution: The process through which species change over time, adapting to their environment through mutation and natural selection.

Food Web / Food Chains: The network of feeding relationships among living organisms. A food chain is a simple line (plant → herbivore → carnivore), while a food web shows how those chains interconnect within an ecosystem.

Fossil Fuels: Coal, oil, and natural gas formed over millions of years from the buried remains of plants and animals. When burned, they release energy—and also gases that warm the planet's climate.

Galaxy / Solar System: A galaxy is an immense collection of stars, planets, gas, and dust bound together by gravity. Our home is the Milky Way. The solar system consists of the Sun and the planets that orbit around it.

Global Warming: The sustained increase in The Earth's average temperature, caused mainly by the accumulation of greenhouse gases due to human activities such as burning fossil fuels and deforestation.

Gravity: The natural force that attracts objects toward one another. Thanks to gravity, planets orbit the Sun and we remain anchored to the ground.

Great Oxidation Event: A period that occurred about 2.4 billion years ago when cyanobacteria began releasing large amounts of oxygen, transforming Earth's atmosphere.

Greenhouse Effect: A natural phenomenon that keeps the planet warm by trapping part of the Sun's heat. When intensified by an excess of gases such as CO_2 or CH_4, it leads to global warming.

Greenhouse Gases: Gases that trap heat in the atmosphere. Examples include carbon dioxide (CO_2), methane (CH_4), and nitrous oxides. In excess, they cause global warming.

Humus / Organic Matter: Humus is the dark, fertile layer formed by fallen leaves, plant remains, and decomposing organisms. Organic matter is that material once transformed—it enriches the soil with nutrients and makes it softer, more porous, and full of life.

Hydrocarbons: Substances made of carbon and hydrogen atoms, forming the basis of fossil fuels such as oil, natural gas, and coal.

LUCA (Last Universal Common Ancestor): The name given to the most ancient common ancestor of all living organisms. LUCA was not a complex being, but a primitive life form from which all others descended.

Magnetic Field: An invisible shield generated by the movement of molten iron in The Earth's core. It protects the planet from solar winds which, without this field, would strip away the atmosphere and make life impossible.

Main Subatomic Particles (protons, neutrons, electrons): The smaller components that make up atoms.

Mantle / Earth's Crust: The crust is Earth's outermost layer, where we live. Beneath it lies the mantle, made of hot, dense rock that, in some places, melts to form magma.

Mass Extinction / Sixth Extinction: A period in which a large number of species disappear within a short geological time span. Five have occurred in the past; today, human activity is driving a sixth.

Microorganisms: Tiny living beings visible only under a microscope— bacteria, archaea, algae, and microscopic fungi.

Mitochondria: Organelles found in nearly all cells. They convert nutrients into chemical energy (ATP) that the cell can use.

Molecule: A group of atoms bonded together. It can be as simple as oxygen (O_2) or as complex as DNA.

Mycelium: A network of fungal filaments (hyphae) that grows underground or within organic matter. It forms part of the living web that connects and nourishes plants.

Mycorrhizae: Associations between fungi and plant roots. The fungi help the plant absorb water and nutrients, and the plant provides sugars produced through photosynthesis. It is one of the oldest and most successful partnerships in the history of life on Earth.

Mycorrhizal Network: The underground web formed by connections between mycorrhizal fungi and the roots of multiple plants. It works like a highway for nutrients and chemical signals, linking trees and other plant species.

Nature-Based Solutions: Actions that harness natural processes to address environmental challenges. Examples include restoring mangroves to protect coastlines, planting forests to capture carbon, and managing living soils to retain water.

Nuclear Fusion: A reaction that occurs in the cores of stars, where hydrogen atoms fuse to form helium, releasing vast amounts of energy.

Nutrient Cycle: The journey of elements such as carbon, nitrogen, and phosphorus through plants, animals, microbes, soil, air, and water. These cycles enable life to renew itself and stay in balance.

Oxygen: A gas essential for almost all life. We breathe it in, and cells use it to release energy.

Photosynthesis: The process by which plants, algae, and some bacteria capture sunlight and transform it into food (sugars), releasing oxygen in the process.

Plankton (Phytoplankton and Zooplankton): A collection of tiny organisms that float in oceans and lakes. Phytoplankton carry out photosynthesis, while zooplankton feed on them.

Pollinators: Animals that carry pollen from one flower to another, allowing plants to reproduce.

Prokaryote / Eukaryote: Prokaryotes are microscopic organisms without a defined nucleus. Eukaryotes have cells with a nucleus and organelles; from them arose plants, animals, and all complex life forms.

Proterozoic: A geological eon that occurred between 2.5 billion and 541 million years ago. During this time, oxygen levels rose and more complex organisms appeared.

Protein: A molecule made of chains of amino acids that performs vital functions in all living beings.

Radiant Energy: Energy that travels in waves, like sunlight. It is the main source of energy for life on Earth.

Regeneration: The ability of ecosystems to recover their balance when given space and care. It also refers to human practices that help nature heal, such as restoring soils, planting trees, or protecting rivers.

Rhizosphere: The zone of soil surrounding plant roots. In this thin layer, bacteria, fungi, and other microbes live together in a web of exchanges that nourishes and protects plants.

Soil Erosion: The process by which water, wind, or poor agricultural practices wear away and carry off the fertile layer of the land.

Soil Fertility: The ability of the land to support plants, thanks to its nutrients, water, and living microorganisms.

Stromatolites: Layered rock formations created by ancient microbial communities, mainly cyanobacteria. They are among the oldest fossils on Earth.

Tipping Point (Point of No Return): A situation in which an ecosystem or the climate undergoes such a major change that it can no longer return to its original state.

Water (cohesion, capillarity, hydrological cycle): Water is the essential molecule for life. Thanks to cohesion, droplets attract one another and can rise through plant stems. Capillarity allows water to climb through very narrow spaces without pumps. Its hydrological cycle is the continuous journey of evaporating, forming clouds, raining, infiltrating the soil, and returning to rivers and seas.

Water Cycle: The continuous movement of water on Earth: it evaporates, forms clouds, falls as rain, infiltrates the soil, flows through rivers, and returns to seas and oceans. This cycle distributes energy and sustains ecosystems.

Complementary References — Parts I To IV

Chapter 1 · A Beginning Of Stars
Sagan, C. (1980). *Cosmos.* Random House.
Chaisson, E. (2001). *Cosmic Evolution: The Rise of Complexity in Nature.* Harvard University Press.
NASA. (2020). *Big Bang.* NASA Science: Astrophysics Division.

Chapter 2 · Earth: An Improbable Oasis
Ward, P., & Brownlee, D. (2000). *Rare Earth: Why Complex Life Is Uncommon in the Universe.* Springer.
Tyson, N. D. (2019). *Astrophysics for People in a Hurry.* W. W. Norton.
National Academies. (2012). *The Limits of Organic Life in Planetary Systems.*

Chapter 3 · The Miracle Of Water
Ball, P. (2008). *H_2O: A Biography of Water.* Weidenfeld & Nicolson.
Chaplin, M. (2006). *Water: Its Importance to Life. Biochemistry and Molecular Biology Education, 34*(3), 165–170.
UNESCO. (2019). *World Water Development Report.*

Chapter 4 · The Layer Of Life
Vernadsky, V. I. (1998, reprint). *The Biosphere.* Copernicus/Springer.
Margulis, L., & Sagan, D. (1995). *What Is Life?* University of California Press.
Lovelock, J. (2000). *Gaia: A New Look at Life on Earth.* Oxford University Press.

Chapter 5 · The Spark Of Life
Hazen, R. (2012). *The Story of Earth.* Viking Penguin.
Lane, N. (2015). *The Vital Question: Energy, Evolution, and the Origins of Complex Life.* W. W. Norton.
Martin, W., & Russell, M. (2003). *On the origin of biochemistry at an alkaline hydrothermal vent. Philosophical Transactions of the Royal Society B, 358*(1429), 59–83.

Chapter 6 · The Great Transformation
Knoll, A. H. (2003). *Life on a Young Planet: The First Three Billion Years of Evolution on Earth.* Princeton University Press.
Schopf, J. W. (1999). *Cradle of Life: The Discovery of Earth's Earliest Fossils.* Princeton University Press.
Falkowski, P. (2015). *Life's Engines: How Microbes Made Earth Habitable.* Princeton University Press.

Chapter 7 · The Web Of Evolution

Margulis, L. (1998). *Symbiotic Planet: A New Look at Evolution.* Basic Books.

Lane, N. (2009). *Life Ascending: The Ten Great Inventions of Evolution.* W. W. Norton.

Maynard Smith, J., & Szathmáry, E. (1995). *The Major Transitions in Evolution.* Oxford University Press.

Chapter 8 · The Dance Of Matter

Odum, E. P. (1971). *Fundamentals of Ecology.* Saunders.

Smil, V. (2017). *Energy and Civilization: A History.* MIT Press.

Falkowski, P., & Raven, J. (2007). *Aquatic Photosynthesis.* Princeton University Press.

Chapter 9 · The Invisible Fabric

Coleman, D. C., Crossley, D. A., & Hendrix, P. F. (2004). *Fundamentals of Soil Ecology.* Academic Press.

Wall, D. H., et al. (2012). *Soil Ecology and Ecosystem Services.* Oxford University Press.

Bonfante, P., & Anca, I.-A. (2009). *Plants, mycorrhizal fungi, and bacteria: A network of interactions. Annual Review of Microbiology, 63,* 363–383.

Chapter 10 · The Unity Of Life

Capra, F. (1996). *The Web of Life: A New Scientific Understanding of Living Systems.* Anchor Books.

Maturana, H., & Varela, F. (1980). *Autopoiesis and Cognition: The Realization of the Living.* D. Reidel.

Wilson, E. O. (2010). *The Diversity of Life.* Belknap Press/Harvard University Press.

Chapter 11 · Listening To The Earth Again

Berry, T. (1999). *The Great Work: Our Way into the Future.* Bell Tower.

Abram, D. (1996). *The Spell of the Sensuous: Perception and Language in a More-than-Human World.* Vintage.

Goodall, J. (2002). *Reason for Hope: A Spiritual Journey.* Warner Books.

Chapter 12 · The Human Dominion

Steffen, W., Crutzen, P., & McNeill, J. (2007). *The Anthropocene: Are humans now overwhelming the great forces of nature? AMBIO, 36*(8), 614–621.

Rockström, J., et al. (2009). *A safe operating space for humanity. Nature, 461,* 472–475.

McNeill, J. R. (2001). *Something New Under the Sun: An Environmental History of the Twentieth-Century World.* W. W. Norton.

Chapter 13 · A Personal Awakening

Kimmerer, R. W. (2013). *Braiding Sweetgrass: Indigenous Wisdom, Scientific Knowledge, and the Teachings of Plants.* Milkweed Editions.

Macy, J., & Johnstone, C. (2012). *Active Hope: How to Face the Mess We're In without Going Crazy*. New World Library.
Kabat-Zinn, J. (2005). *Coming to Our Senses: Healing Ourselves and the World through Mindfulness*. Hyperion.

Chapter 14 · Realistic Hope
Hawken, P. (2017). *Drawdown: The Most Comprehensive Plan Ever Proposed to Reverse Global Warming*. Penguin.
Klein, N. (2014). *This Changes Everything: Capitalism vs. the Climate*. Simon & Schuster.
IPCC. (2021). *Climate Change 2021: The Physical Science Basis*. Intergovernmental Panel on Climate Change.

Chapter 15 · Two Paths Before Us
Wallace-Wells, D. (2019). *The Uninhabitable Earth: Life After Warming*. Tim Duggan Books.
Lenton, T. M., & Latour, B. (2018). *Gaia 2.0*. Science, 361(6407), 1066–1068.
Ripple, W. J., et al. (2020). *World scientists' warning of a climate emergency*. BioScience, 70(1), 8–12.

Chapter 16 · The Threshold
Shiva, V. (2005). *Earth Democracy: Justice, Sustainability, and Peace*. South End Press.
Latour, B. (2017). *Facing Gaia: Eight Lectures on the New Climatic Regime*. Polity Press.
Meadows, D. H., Meadows, D. L., & Randers, J. (2004). *Limits to Growth: The 30-Year Update*. Chelsea Green.
Rockström, J., et al. (2009). *A safe operating space for humanity*. Nature, 461, 472–475.
Steffen, W., et al. (2015). *The trajectory of the Anthropocene*. The Anthropocene Review, 2(1), 81–98.
Lewis, S. L., & Maslin, M. (2018). *The Human Planet: How We Created the Anthropocene*. Penguin / Random House.

Chapter 17 · When We Opened The Vault Of The Past
Lovelock, J. (2000). *Gaia: A New Look at Life on Earth*. Oxford University Press.
Margulis, L., & Sagan, D. (1995). *What Is Life?* University of California Press.
Berner, R. A. (2004). *The Phanerozoic Carbon Cycle: CO_2 and O_2 Through Time*. Oxford University Press.
Falkowski, P. (2015). *Life's Engines: How Microbes Made Earth Habitable*. Princeton University Press.
Odum, H. T. (1996). *Environmental Accounting: Emergy and Environmental Decision Making*. Wiley.
Smil, V. (2017). *Energy and Civilization: A History*. MIT Press.

Chapter 18 · Returning To The "We"

Margulis, L., & Sagan, D. (1995). *What Is Life?* University of California Press.

Abram, D. (1996). *The Spell of the Sensuous: Perception and Language in a More-than-Human World.* Pantheon Books.

Berry, T. (1999). *The Great Work: Our Way into the Future.* Bell Tower.

West, G. (2017). *Scale: The Universal Laws of Growth, Innovation, Sustainability, and the Pace of Life.* Penguin Press.

Lovelock, J. (2000). *Gaia: A New Look at Life on Earth.* Oxford University Press.

Chapter 19 · The Pact With Life

Berry, T. (1999). *The Great Work: Our Way into the Future.* Bell Tower.

Shiva, V. (2005). *Earth Democracy: Justice, Sustainability, and Peace.* South End Press.

Daly, H. (1996). *Beyond Growth: The Economics of Sustainable Development.* Beacon Press.

Lovelock, J. (2000). *Gaia: A New Look at Life on Earth.* Oxford University Press.

Ward, P. D., & Brownlee, D. (2000). *Rare Earth: Why Complex Life Is Uncommon in the Universe.* Springer.

Chapter 20 · The Continuity We Are

Malm, A. (2016). *Fossil Capital: The Rise of Steam Power and the Roots of Global Warming.* Verso.

Georgescu-Roegen, N. (1971). *The Entropy Law and the Economic Process.* Harvard University Press.

Daly, H. (1996). *Beyond Growth: The Economics of Sustainable Development.* Beacon Press.

IPCC. (2021). *Sixth Assessment Report – Working Group I: The Physical Science Basis.* Intergovernmental Panel on Climate Change.

The Earth's Mirror: Evidence And Data

This book has sought to awaken awe, gratitude, and respect for The Earth.

But upon reaching this point, I felt there was one last stop needed: to show the quantitative evidence—without embellishment, without metaphor. Because poetry helps us feel, but data demands our attention.

What the reader will find in these pages contains no exaggeration. Every fact presented is supported by scientific research and international reports.

Science works as a community. Teams of specialists in different fields—often from several countries—collaborate to collect, analyze, and compare data. Before results are published, they must undergo a process of critical peer review—by at least two or three independent experts—who evaluate the quality and reliability of the research. Only then does it become accepted knowledge within the scientific community.

Although there can always be isolated exceptions driven by economic or political interests, the vast majority of the thousands of scientists around the world share a single purpose: to seek truth, verify data, and reproduce results that can guide collective decisions. Thanks to this rigorous and collaborative work, we now have reports and graphs that clearly reveal the real state of our planet.

That is what this section presents: solid findings, built with the seriousness that defines scientific inquiry. And although the numbers may seem cold, each graph tells a story—the story of how, in just a few decades, we have transformed the forests, oceans, soils, air, and even the space above our heads.

At the same time, this section is not meant to dwell in sadness or fear. It is meant to open both eyes and heart. Because the planet has no borders: what happens in one place echoes everywhere, and we all must act.

No country can close its eyes to escape the problem. The greatest injustice occurs when those least responsible for the damage are the ones who suffer its consequences the most.

While those who pollute the most have the resources to protect themselves—or believe they can—with bunkers and underground cities. But those are not ways of living fit for anyone.

This is a call to awareness—to recognize that we cannot trade a paradise of open air, clean water, and abundant life for an artificial confinement. Yes, these numbers are stark. But they are also the compass reminding us what is at stake, urging us to act with both urgency and hope.

1. Energy and Emissions

The recent history of humankind can be told with a single match: we struck it to ignite fossil fuels—and we have not stopped since. Coal, oil, and natural gas became the lifeblood of modern life. They are present in more places than we imagine.

Every time we start a car to drive a few blocks, open the refrigerator that keeps everything cold day and night, or breathe the air conditioning that cools our home, fossil fuels are behind it. The cement used to build our schools, hospitals, cinemas, and homes releases tons of CO_2 during production. Every airplane trip—and even more so in private jets—runs on kerosene that ends up in the atmosphere as greenhouse gases. Thousands of boats and marine engines hum each day along our rivers and coasts, adding more emissions.

At the planetary scale, fossil fuels are the main source of CO_2: it is estimated that every year more than 35 gigatons (Gt) of carbon dioxide are released from combustion alone, and about 35% of that comes from oil.

Every gas tank, every sack of cement, every flight, and every air conditioner turned on is part of that sum that is already altering our planet's climate.

And let's think further: plastic. That little bag we're given to carry something for three blocks—made from oil, shipped by sea and truck, and discarded within minutes—is only one example of how every daily action adds emissions and leaves an ecological footprint.

The data never lie: what Earth formed and stored underground over millions of years—in the form of coal, oil, and gas—we are extracting and burning in just a few generations. Transportation alone consumes almost half of all the oil produced worldwide **(Figure 1)**, which means every kilometer traveled carries a measurable consequence.

Global uses of oil in 2015

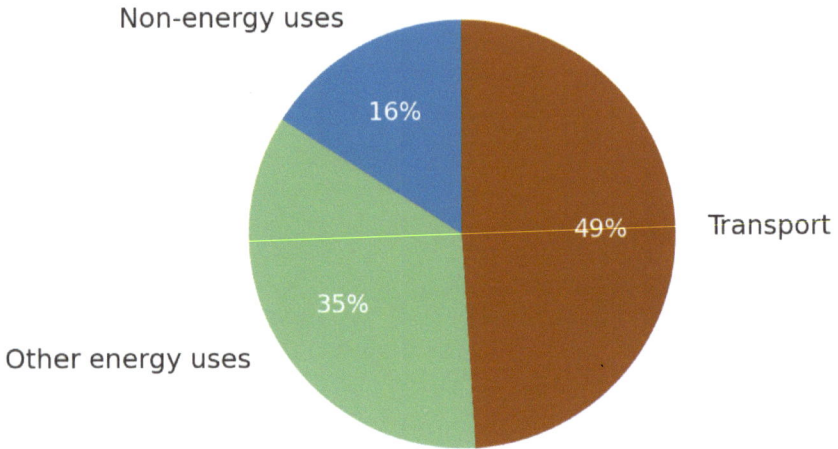

Non-energy uses

16%

Transport

49%

35%

Other energy uses

Figure 1. Global uses of petroleum in 2015.
Source: IEA, BP, EIA, and WEC (2016).

More than sixty percent of the world's great fortunes come directly or indirectly from oil, gas, or the industries that depend on them—transportation, plastics, fertilizers, and finance. The five largest oil companies—ExxonMobil, Chevron, Shell, BP, and TotalEnergies—recorded combined profits in 2023 exceeding 200 billion dollars, even as the planet faced unprecedented fires, droughts, and heat waves.

The wealth of the twenty-first century has been built, to a great extent, on the very same fuels that are destabilizing the climate that sustains life.

2. Agriculture and Food

Part A – Pesticides and Glyphosate

To increase harvests and respond to the growing demand for food, modern agriculture became dependent on pesticides: herbicides, insecticides, and fungicides. Since the mid-20th century, their use has only increased. What once was minimal use has turned into a large-scale, everyday practice.

In short: every spotless tomato, every perfect lettuce leaf, every "weed-free" lawn is usually sustained by the application of synthetic chemicals—millions of liters falling on soils, waters, and food. Figure 2 shows the global distribution of pesticides used in agriculture. Herbicides are the most widely used, followed by insecticides, fungicides, and bactericides.

Global use of pesticides in agriculture (2019)

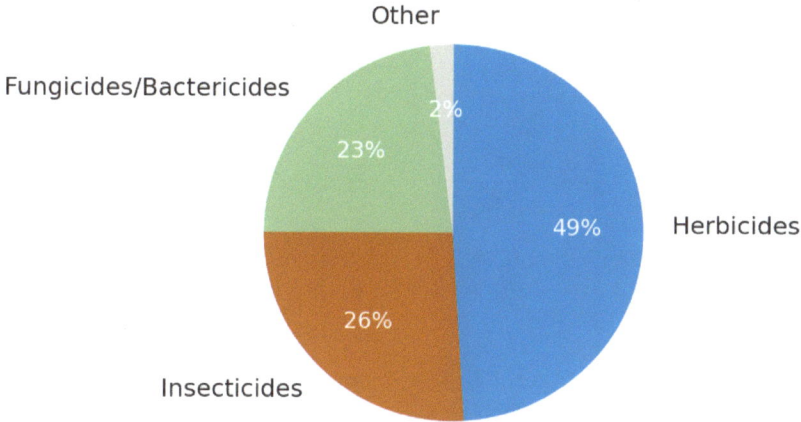

Figure 2. Global pesticide use in agriculture (2019).
Source: FAO (2019) and WHO (2020).

Glyphosate is the clearest example. For a small farmer, it could once be a punctual tool—used carefully to clear small areas without harming the crop, reducing competition from wild plants, and slightly improving yield.

But in the 1990s, with the arrival of genetically modified crops engineered to tolerate it, everything changed. Suddenly, there were no limits: spraying could be done without fear. The cultivated plant would survive.

That opened the door to massive, repeated use—year after year—reaching nearly 200 million hectares. Since the introduction of glyphosate-tolerant genetically modified crops in the 1990s, glyphosate use has grown explosively (Figure 3).

Today, hundreds of millions of kilograms of this herbicide are applied worldwide every year.

Global use of glyphosate and GM-tolerant crops (1974–2020)

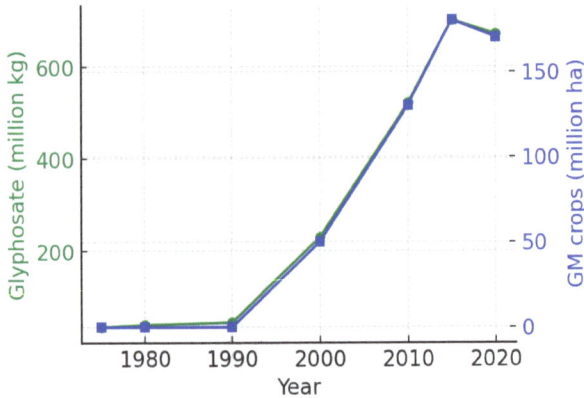

Figure 3. Global glyphosate use and GM-tolerant crops (1974–2020).
Source: Benbrook (2016); FAO (n.d.); ISAAA (2020).

The graph speaks for itself: before 1996, glyphosate was used in relatively small quantities. With the introduction of resistant crops, the curve skyrocketed. This was no coincidence—the technology promoted as a solution to low yields opened a path to dependence on glyphosate.

This herbicide doesn't stay on the farms: we find it in soybean oil used for cooking, in the corn that becomes tortillas, and in our yards, where it's often sprayed without a second thought about its side effects. What once seemed like a useful agricultural chemical has now become present in our food—with potential consequences for human health.

Part B – Meat and GM Crops

But the impact of modern agriculture is not measured only by the presence of agrochemicals. It is also found in what we choose to eat.

Global beef consumption has surged in recent decades—and with it, pressure on land, water, and climate. Every steak or hamburger represents more than the animal itself. It also represents the fields of corn and soy grown to feed it—mostly genetically modified crops sprayed with glyphosate—the forests cleared to make room for pastures, and the methane released by that livestock into the atmosphere.

In fact, global beef consumption has doubled since 1960 and continues to rise, with projections reaching nearly 70 million tons by 2030 (Figure 4).

When we sit in front of a hamburger, we rarely think of deforestation in the Amazon or the tons of pesticides used to produce it. Yet the chain is there. Every consumer choice adds up: one more portion of meat, one less forest.

The good news is that we have options. This isn't about eliminating food overnight, but about recognizing that every reduction in industrial meat consumption, every choice of local and agroecological products, every conscious decision to support different ways of producing food, multiplies the positive impact.

Global beef consumption (1961–2030)

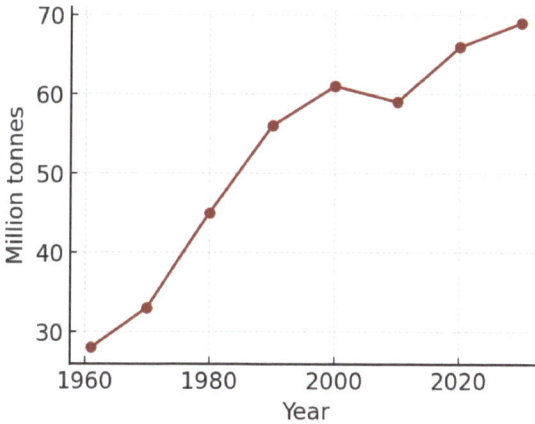

Figure 4. Global beef consumption (1961–2030).
Source: Our World in Data, based on Poore & Nemecek (2018).

3. Forests

The world's forests are the lungs and the heart of The Earth. Yet in recent decades, we have treated them as if they were endless warehouses of wood, land, and space.

According to the FAO and Global Forest Watch, the global rate of deforestation peaked in the 1980s and 1990s. And although, in global numbers, it may appear to have "slowed," what we continue to lose today is the most precious of all: the planet's remaining primary tropical forests—those that have never been logged and hold the greatest concentration of biodiversity and carbon on Earth.

What remains standing is the most sacred—and still, we are cutting it down.

Where does this show up in our daily lives?

- In the paper napkin we use once and throw away, when we once used cloth napkins that were washed and reused.

• In the lunch boxes and cloth bags that were replaced by disposable cartons and plastic bags. Once they were washed and lasted for years; now everything is carried in packaging that lasts only minutes before becoming trash.

• In the cardboard boxes that wrap and ship food from one country to another—millions of tons of cardboard that last only as long as the trip, but mean entire forests turned into waste.

• In mass-produced furniture made from cheap wood.

• In cookies, chocolates, or cosmetics containing palm oil—one of the main drivers of deforestation in Southeast Asia.

• In cheap beef, since large areas of the Amazon rainforest are cleared to create pastures or to plant soy that later becomes livestock feed.

• In agricultural lands rented in some countries to grow feed for the livestock of others.

Global deforestation remains alarming, even though the annual rate has declined since 1990—from around 16 million hectares per year in the 1990s to about 10 million hectares per year in the past decade (Figure 5).

Annual global deforestation (1990–2020)

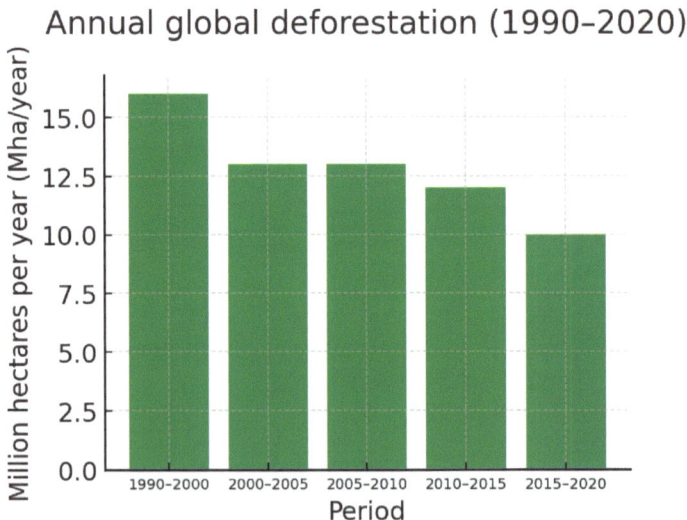

Figure 5. Annual global deforestation (1990–2020).
Source: FAO (2020); FAOSTAT (n.d.).

However, the estimates are not uniform. While FAO reports a gradual decline in *net* deforestation, data from Global Forest Watch show a much larger and more fluctuating loss of tree cover (Figure 6).

It's essential to understand the difference: when a forest is cut down, we don't just lose "a hundred trees." We destroy an entire ecosystem, including the invisible life beneath the soil—fungi, microbes, roots—that took millions of years to form.

Planting a hundred trees somewhere else will not replace that complex network that sustains soil fertility, biodiversity, and climate balance.

Net deforestation (FAO) vs. tree cover loss (GFW, 1990–2023)

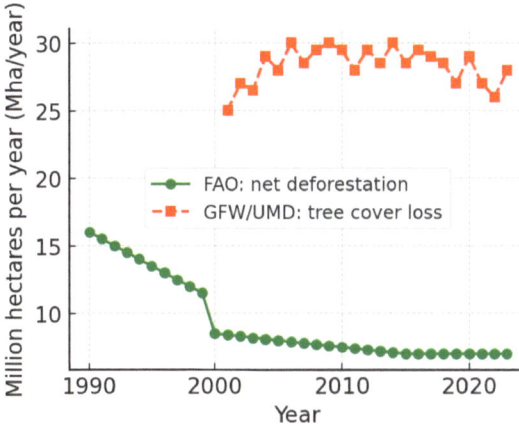

Figure 6. Net deforestation vs. tree-cover loss (1990–2023).
Source: FAO (2020); Global Forest Watch (2001–2023).

Every paper napkin, every discarded cardboard box, every cheap piece of furniture we buy without thinking adds to this tally. We always have the choice to tilt the balance the other way: return to cloth napkins, rescue the reusable lunch box, choose certified wood, reduce paper waste, and support forest-conservation projects.

In the end, what still stands is what sustains the air we breathe. If we care for it, we care for ourselves.

4. Oceans

For centuries, the sea gave us the illusion of being infinite. Fish were plentiful; tuna seemed inexhaustible. But industrial fishing changed everything: giant nets, factory ships, satellites that leave nothing unseen.

Since 1950, marine catches have risen steadily until reaching a plateau of around 80 million tons per year.

It's clear we have already taken from the ocean nearly all it could give.

What happens at home ends up on your plate. In every can of tuna, every frozen fillet, every piece of sushi — and even in the shrimp or lobster ordered as a luxury — the ocean's story is written.

Cod, once considered common people's food and now sold at luxury prices, became scarce because it was overfished.

Global marine catches grew rapidly during the second half of the 20th century, but since the 1990s they have remained on a plateau of about 80 million tons per year, reflecting the maximum pressure that the oceans can sustain (Figure 7).

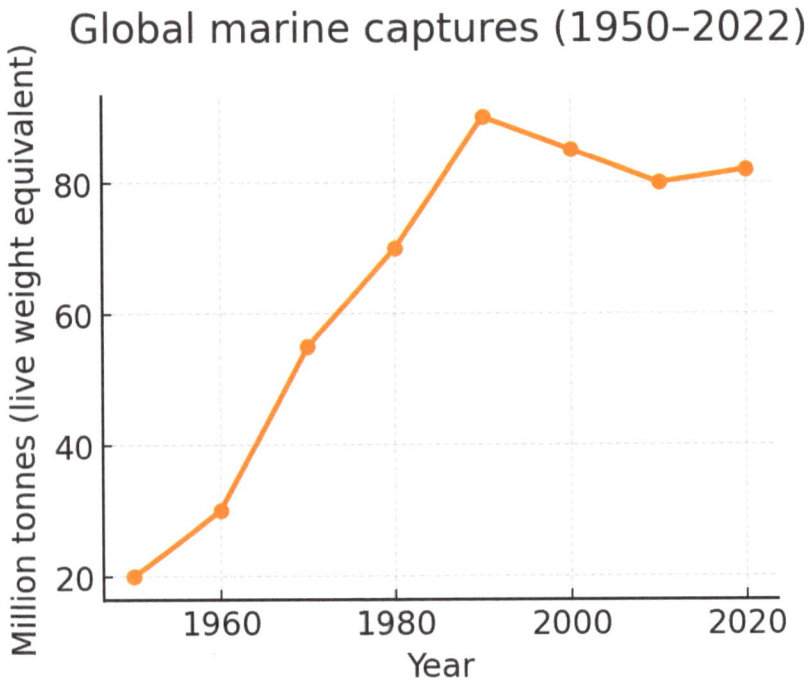

Figure 7. Global marine catches (1950–2022).
Source: FAO (2022); FAOSTAT (n.d.).

And when fishing alone was no longer enough, we took another step: aquaculture. In 2022, farmed production surpassed wild-capture fisheries for the first time. As marine catches have stagnated since the 1990s, aquaculture has skyrocketed—recently overtaking wild fisheries as the main global source of aquatic food (Figure 8).

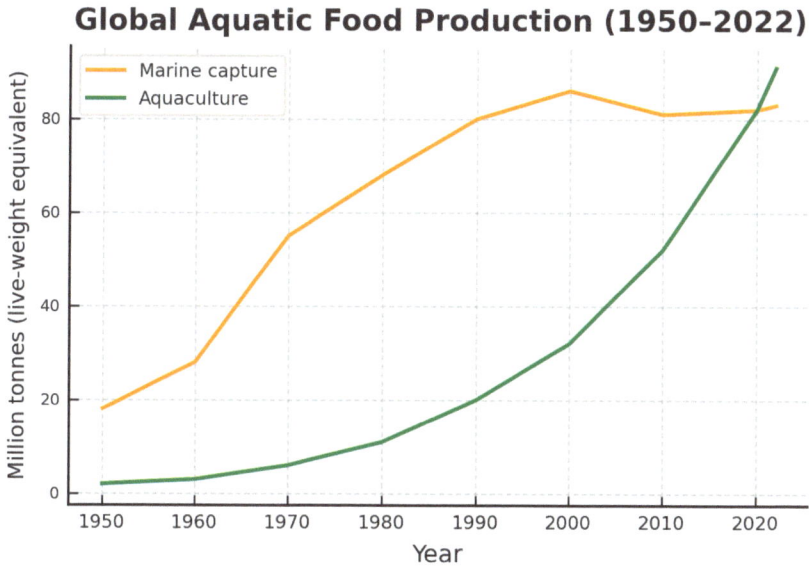

Figure 8. Global aquatic food production (1950–2022).
Source: FAO (2022); FAOSTAT (n.d.).

The Ocean's Other Lung: Corals Are Suffocating Too

Coral reefs are like the tropical forests of the sea — shelters for fish, natural barriers against storms, and generators of life. But today, they are dying.

In Puerto Rico, a mass bleaching event in 2005 wiped out up to 63 % of corals in areas such as Desecheo and Mona. Since 2019, diseases like Stony Coral Tissue Loss Disease have been devouring entire species. And what happens in La Parguera and Vieques matters far beyond tourism — the lens through which government action is often justified. Those reefs absorb carbon, sustain fisheries, and protect coastal communities from storms.

In Australia, the situation is equally alarming. A record-breaking marine heatwave in 2024 caused the largest annual decline in coral cover in the last 39 years: the northern section of the reef lost about 25 %, the central section nearly 14 %, and the southern section almost a third of its living coral. Drone-based surveys near Lizard Island revealed that 96 % of corals died within just three months, marking one of the highest mortality rates ever recorded worldwide.

According to a recent report by 160 scientists from 23 countries, warm-water coral reefs have now crossed a climate point of no return in 2025—meaning that even if emissions dropped to zero tomorrow, they could no longer recover on a global scale. The authors warn that this coral collapse represents the first confirmed tipping point in The Earth system, carrying the risk of a domino effect toward other critical ecosystems.

Mass Coral Bleaching Event in Puerto Rico (2005)

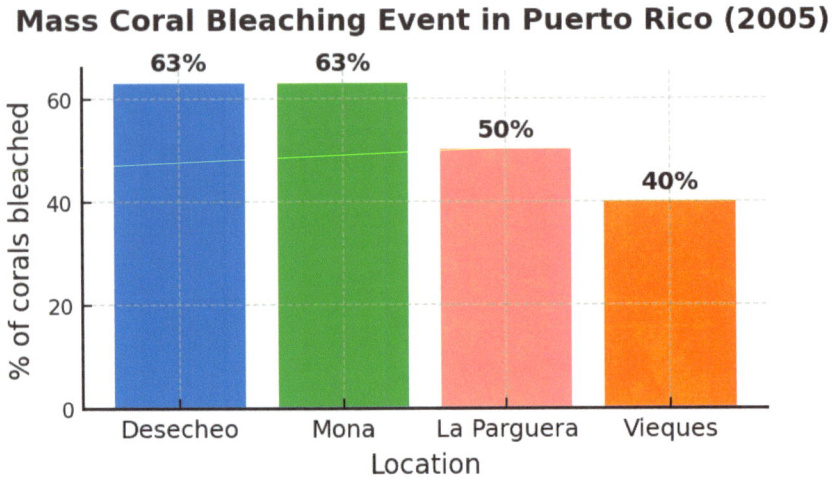

Figure 9. Coral bleaching in Puerto Rico (2005).
Source: NOAA (2005); Puerto Rico Coral Reef Monitoring Program.

5. Plastics

Plastic was invented as a wonder material: light, strong, moldable, and almost eternal. It sounds ideal for making things that last... yet we mostly use it for the opposite: objects that last minutes.

In 2019, the world produced around 460 million tons of plastic, and nearly half of it was single-use—bags, bottles, and packaging designed to be discarded within minutes, yet destined to remain on the planet for centuries (Figure 10).

If in the Energy section we saw how oil fuels cars, factories, and airplanes, here we see it transformed into something even closer: the objects of our daily lives.

Global Plastic Production (2019)

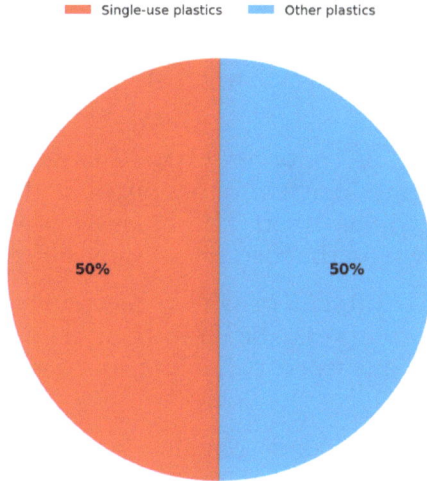

Figure 10. Global plastic production (2019).
Source: Geyer et al. (2017); Our World in Data (2019).

Where does it appear in our everyday life?
- In the water bottle we buy out of convenience, even when we have water at home.
- In the straw we use for ten minutes and throw away.
- In disposable cutlery and plates used for a party or takeout meal.
- In the unnecessary wrapping that covers fruit, cookies, or products that already had a natural peel.
- In the plastic grocery bags that carry what we walk only a few blocks with.

And those fragments are everywhere:
in table salt,
in the water we drink,
in the fish we eat,
in the air we breathe,
and even in the placenta of babies yet to be born.

What we use for seconds stays on the planet for centuries—and it is already entering our own bodies.

The good news is that the solutions are also found in everyday life.

Carrying our own water bottle.

Saying no to the straw.

Using cloth bags.

Returning to cloth napkins and reusable containers like the fiambrera.

Small steps that, multiplied by millions of people, reduce the mountain of plastic we are creating.

But the most concerning part is not what we see—it's what we don't. Microplastics.

Every plastic bag, bottle, or container exposed to sunlight and time breaks into tiny fragments. Each time we wash synthetic clothing, thousands of microscopic fibers flow out with the water. Each tire that wears down on the road releases invisible plastic dust.

Microplastics are already present in salt, in the water we drink, and even in the placenta of babies yet to be born. Their main sources are not just the degradation of discarded packaging, but also other hidden pathways (Figure 11).

In short: what we use for seconds remains for centuries, and it is already entering our own bodies.

Global use of pesticides in agriculture (2019)

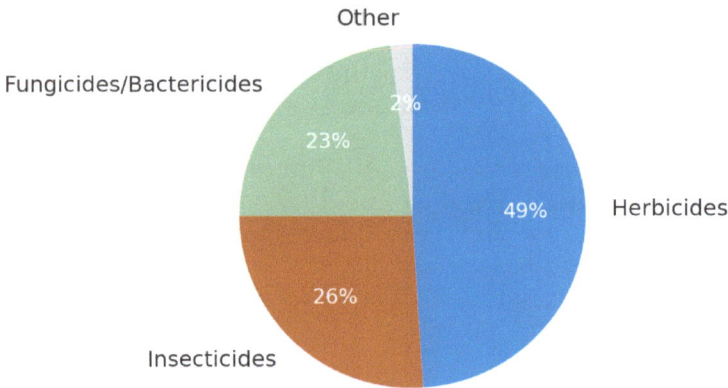

Figure 11. Sources of microplastics in the environment.
Source: Boucher & Friot (2017); UNEP (2019).

6. Minerals and Sand

Technology has become our digital bloodstream. Like an invisible serum, it flows through our lives — in cell phones, computers, and tablets. All these devices connect us, inform us, and entertain us. But we rarely think about what lies behind the screen—and what it takes for it to work.

Every sleek, lightweight device hides kilograms of minerals torn from

The Earth: copper, nickel, lithium, cobalt, and rare earth elements — all essential for the batteries, chips, and magnets that sustain this connected life. Their extraction has skyrocketed since the mid-20th century, destroying mountains across Latin America, Africa, and Asia (Figure 12).

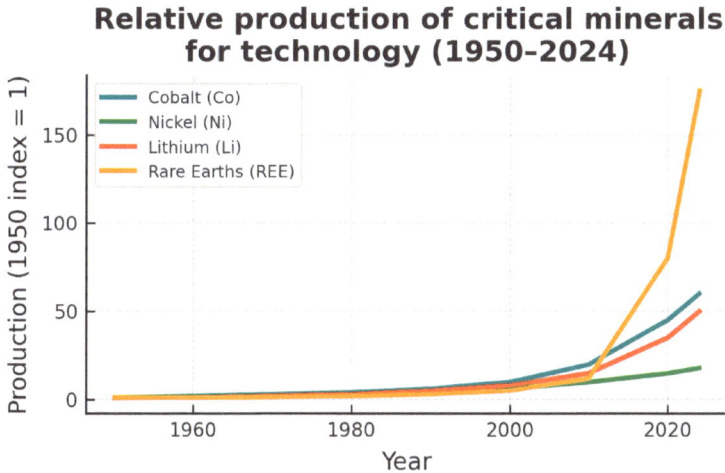

Figure 12. Growth in the extraction of critical minerals for technology. *Source: USGS (various years); World Bank (2020).*

To make a phone that fits in your hand work, The Earth must be excavated as never before in human history.

- Where do we see it in daily life?
  - In the cell phone we carry everywhere and replace so often.
  - In the laptop or tablet from which we work, study, or read these very words.
  - In the electric car that promises to be "green," yet depends on lithium, nickel, and cobalt from massive open-pit mines.
  - In flat screens and wireless earbuds full of rare earth elements.
  - Even in the refrigerator, washing machine, and microwave that we can no longer imagine living without.

The most extracted resource on the planet is not oil, nor coal, nor even copper — it's sand. We use it to make cement, glass, asphalt, highways, and entire cities. By extracting it, we irreversibly alter river deltas and, as a result, our coastlines.

In the Energy section we saw how cement releases CO_2 during production; here we see its other side: its main ingredient — sand. Every building and every road we construct carries tons of it, and global demand now exceeds what rivers can replenish. The extraction of sand and gravel has surged in recent decades, reaching over 50 gigatons per year by 2024 — an unsustainable level for rivers, beaches, and freshwater ecosystems (Figure 13).

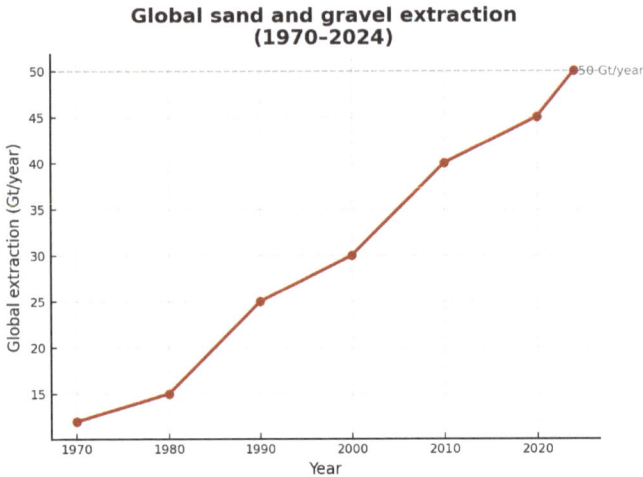

Figure 13. Global sand and gravel extraction (1970–2024). *Source: UNEP (2019); Torres et al. (2021).*

We live immersed in an invisible web that seems to connect us "to everything with nothing," without realizing that this connection feeds on holes growing deeper in the earth and rivers that once ran alive, now dry.

The so-called green energies —electric cars, solar panels, wind turbines— also come at a cost. They are not perfectly clean solutions, because they depend on this same intensive mining. But there is one crucial difference: continuing to burn oil and coal will lead us to collapse within a few decades, whereas these technologies, though imperfect, buy us vital time —time to rethink, to transform the model, to renew our pact with The Earth.

If we fail to do that —if we simply replace one resource with another without changing the logic beneath it—we will end up exhausting even what we call green. But if we use this pause to truly change, we still have a chance to sustain life.

7. Invisible Microorganisms

Beneath our feet, at the bottom of the sea, in rivers and caves, there lives a world we barely know: the underground microorganisms. They are bacteria, archaea, fungi, and viruses we cannot see, yet they form the foundation of life's cycles on Earth.

Science has described barely 15,000 species of subsurface microbes, but estimates suggest there could be over a million (Figure 14). In truth, what we know is just the tip of the iceberg —the invisible life waiting to be discovered is overwhelming.

What do these tiny beings do?
Much more than we can imagine:

•

- They transform nutrients so plants can grow.
- They recycle dead organic matter, returning fertility to the soil.
- In the deep ocean, they convert chemical compounds into energy, sustaining ecosystems where sunlight never reaches.
- They balance gases such as carbon dioxide and nitrogen.

What's most surprising is that these unseen microbes are also part of our daily lives, even if we never notice them:

•

- In garden soil, enabling corn, coffee, or avocados to grow.
- In the beds of rivers and lakes, where they purify water.
- In cheeses, breads, and ferments, all dependent on domesticated microbes.

And yet, with every pesticide, every oil spill, every excess of fertilizer, we are disrupting that hidden world.

We are poisoning something we do not fully understand —
something that sustains life itself.

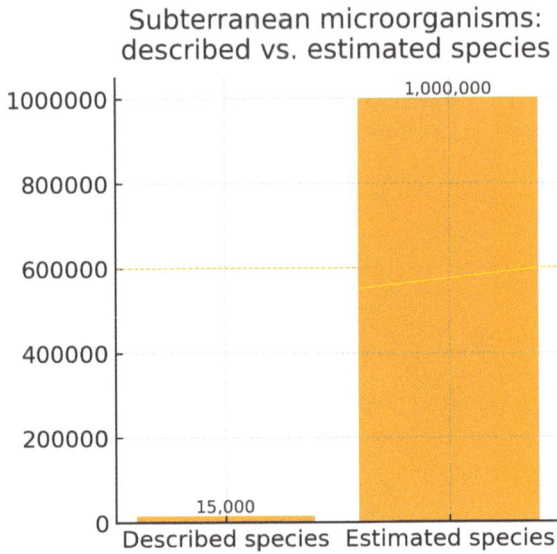

Figure 14. Subsurface microorganism species.
Source: Hug et al. (2016); Whitman et al. (1998).

Killing invisible microbes is like breaking a house's foundation without realizing it. The walls might hold for a while, but sooner or later, everything collapses.

The good news is that we can also protect them —by using fewer synthetic toxins, regenerating soils with compost and green manures, caring for rivers, and recognizing that the invisible is as vital as the visible. Microbes are silent allies, and recognizing their importance is a step toward a deeper pact with The Earth.

8. Satellites and Space Debris

What goes up must come down.

That is one of nature's simplest laws. And yet, we have filled the space surrounding Earth as if it were an eternal landfill.

Since the year 2000, the number of satellites in orbit has exploded— today there are over 9,000 active, plus thousands more that are dead, abandoned, or broken into fragments drifting through space.

Just as we have littered the planet's surface, we are now cluttering its atmosphere.

Where do we see it in daily life?
 • In the GPS we use to reach any destination.
 • In the satellite television we watch at home.
 • In the cell phone that depends on space-based communication.
 • In weather forecasts that rely on satellite data.
 • In bank transactions and internet signals that travel through space.

The number of satellites orbiting Earth has surged in recent years—from just a few thousand in 2015 to over 12,000 projected for 2025—reflecting the boom in private constellations such as Starlink (Figure 15).

But it's not only vital services. Each technological whim, each unnecessary gadget, each invention that solves nothing real also depends on satellites—and, in the end, adds more junk to the sky. What seems like a harmless indulgence becomes part of a global problem.

Number of operational satellites
in Earth orbit (2000–2025)

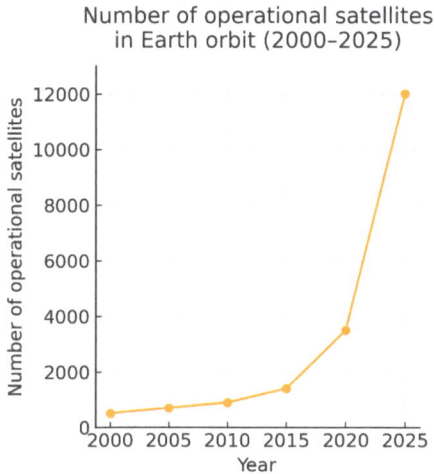

Figure 15. Satellites in Earth orbit (2000–2025).
Source: Progressive Policy Institute; Union of Concerned Scientists; TS2 Space; McDowell (various years).

And the problem is simple: everything that goes up eventually comes down. It's already happening. Pieces of satellites and rocket fragments are falling back to Earth. Most burn up in the atmosphere—but some reach the ground intact, becoming dangerous debris. And the most worrying part is that we can't predict exactly where they'll land. It could be in the ocean, in a desert... or over a city.

We have turned even the sky into a dumping ground,
and now that waste is beginning to fall back on us.

The good news is that there are already proposals to clean up Earth's orbit, and new regulations requiring old satellites to be de-orbited. But unless we act soon, the cloud of debris will grow so large that it could threaten not only space exploration, but also the communications and services we depend on every day.

9. Ultra-Billionaires and Inequality

In the past two decades, the number of ultra-billionaires in the world has skyrocketed. In 2000, there were just a few hundred; today, there are over 2,600 people whose fortunes exceed the gross domestic product of entire countries. Meanwhile, half of the world's population survives on less than six dollars a day.

A handful live in excess, while billions barely survive.

Where does this appear in our daily lives?

> • In the digital platforms we use every day —our phones, social media, and online shopping— that generate gigantic fortunes for a few owners
>
> • In the luxury underground bunkers built by multimillionaires to "survive" the very climate crisis they helped create.
>
> • In so-called "private sustainable cities," promising comfort for a privileged few while most of humanity faces extreme heat, droughts, or floods.
>
> • In the absurd luxuries: a yacht the size of a building, a private jet for one person, a mansion that consumes in a single day the energy a common household uses in a year.

While millions face poverty and environmental degradation, the number of ultra-billionaires in the world has tripled in the past two decades, surpassing 2,700 in 2024 (Figure 16).

The problem is not that wealth exists, but the disproportion: while some accumulate more than they could spend in a hundred lifetimes, others lack the most basic necessities —clean water, sufficient food, medical care, or a safe roof. And that concentration of economic power also means concentration of political power, which slows collective decisions needed to care for the planet.

Number of ultra-billionaires in the world (2000–2024)

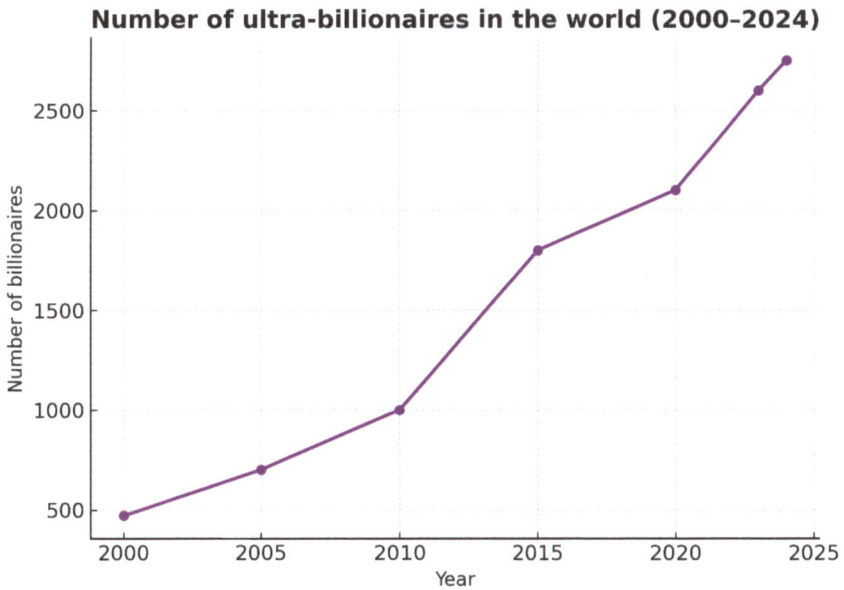

Figure 16. Ultra-billionaires in the world (2000–2024).
Source: Forbes (2024); Statista (2024).

While the ultra-rich seek to save themselves in bunkers, the rest of the world lives in the same common home —with hot walls and cracked ceilings. There is no individual escape: what is destroyed here affects us all.

Although great fortunes are responsible for a disproportionate share of emissions and environmental pressure, a shift is beginning to emerge within parts of this elite. Recent reports show that philanthropy directed toward climate and nature conservation has grown in recent years (ClimateWorks Foundation, Funding Trends 2023), reflecting a certain awareness of the planet's urgency.

While these contributions remain insufficient compared to the magnitude of the challenge, they demonstrate that capital concentrated in few hands can be directed not only toward personal accumulation, but also toward improving life on Earth and ensuring a habitable future for coming generations.

This is not merely a reproach; it is also an invitation: the wealth that weighs so heavily upon The Earth could, if guided with wisdom, help sustain it.

10. War and Competition

Since the year 2000, global military spending has skyrocketed, now exceeding 2.7 trillion dollars per year —the highest figure in decades.

While environmental and social crises intensify, military budgets continue to grow unchecked, reaching over 2.6 trillion dollars in 2023 (Figure 17).

Global military spending (2000-2023)

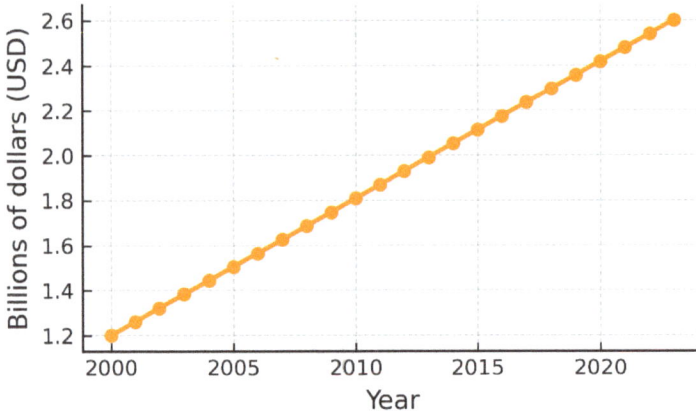

Figure 17. Global military spending (2000–2023).
Source: SIPRI (2023); World Bank (2023).

We often hear that "there is no money" to confront climate change, to protect forests, or to feed the hungry. Yet there always seems to be money —and plenty of it—to produce bombs, fighter jets, and missiles.

But the problem isn't only economic. War costs human lives: millions of dead, wounded, and displaced. And it also costs non-human lives: burned forests, polluted rivers, animals exterminated in silence.

Manufacturing weapons damages the planet even before they are used. Mining metals for tanks and missiles, producing gunpowder, plastics, uranium—all of it leaves deep environmental scars. And when those weapons are used, the effects multiply: oil spills during Middle East wars, contaminated soils in Vietnam, radiation in places like Chernobyl and Fukushima, and entire landscapes turned into sterile deserts.

And from these same damaged lands often come the foods we eat.

The contrast becomes even more staggering when we compare what we spend on destruction with what we would need to spend to restore The Earth.

The cost of restoration is large—but far smaller than the cost of losing nature itself. According to World Bank estimates, the investment required would be about 500 billion dollars per year—a fraction of global military spending—yet enough to save entire ecosystems (Figure 18).

A recent report revealed that more than 61 million tons of debris, much of it contaminated with asbestos and unexploded munitions, now covers the Gaza Strip after years of bombardment, causing severe water and soil pollution and widespread respiratory illness.

It is also estimated that over 80% of agricultural land has been destroyed, that less than 10% of hazardous waste is safely disposed of, and that the direct dumping of untreated sewage into the sea has turned this crisis into what experts describe as a possible "ecocide."

Estimated annual investment for ecosystem restoration

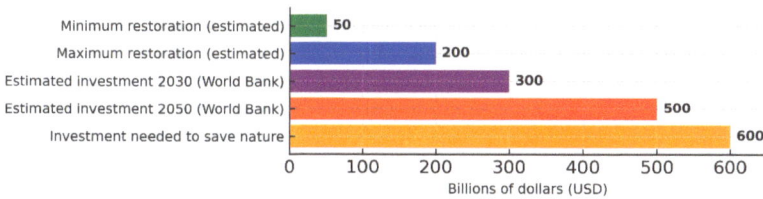

Minimum restoration (estimated)	50
Maximum restoration (estimated)	200
Estimated investment 2030 (World Bank)	300
Estimated investment 2050 (World Bank)	500
Investment needed to save nature	600

Billions of dollars (USD)

Figure 18. Investment in ecosystem restoration.
Source: World Bank (2022); United Nations (2021).

With just a fraction of what we spend on weapons, we could cover the cost of reforesting degraded lands, protecting biodiversity, and restoring ecosystems vital to life.

While the world pours over two trillion dollars into war every year, restoring the planet would require between 31 and 210 billion dollars annually—and even the highest estimate is tiny compared to military spending.

It is astonishing to realize that we spend on destroying ourselves what could rebuild our common home.

Nature has never survived through absolute competition. Life endures through collaboration— the fungi that feed the trees, the corals that shelter the fish, the microbes that enrich the soil.

If we devoted the same energy and resources that we now invest in destroying each other to caring for The Earth and for one another, we would have a healthy and just planet.

II. The Poles

The melting of Earth's poles is one of the clearest and most alarming indicators of climate change. Scientific evidence shows that we are accelerating a process with profound consequences for both the planet and humanity.

In the Arctic, the minimum sea ice extent recorded each September has declined by 12.4% per decade since 1979. The 1981–2010 average was 6.3 million km², while in 2012 it reached a record low of 3.39 million km². By 2024, the value stood at 4.28 million km², confirming that the trend remains far below historical levels (Figure 19).

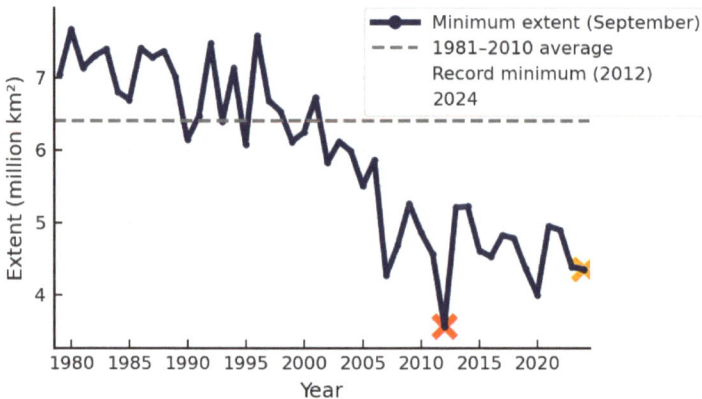

Decline in minimum Arctic sea ice extent (1979-2024)

Figure 19. Minimum Arctic sea ice extent (1979–2024).
Source: NSIDC (2024); NASA Climate Change Vital Signs (2024).

In the Antarctic, the loss of ice mass is equally alarming. Since 2002, an average of 150 gigatons of ice per year has been lost —a total of more than 3,200 gigatons in two decades. Although this represents only 0.012% of Antarctica's total ice mass, it already accounts for over one centimeter of global sea level rise, and the most concerning fact is that the rate of loss is accelerating (Figure 20).

Antarctic ice mass loss (2002-2023)

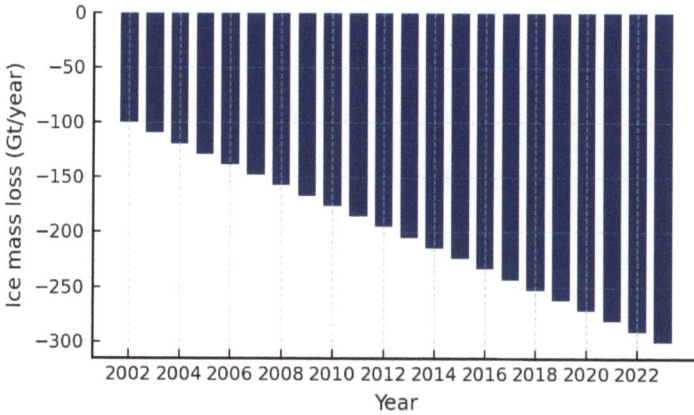

Figure 20. Antarctic ice mass loss (2002–2023).
Source: NASA GRACE/GRACE-FO (2002–2023); IMBIE (2018); NASA Climate Change Vital Signs (2024).

As a result, sea levels are currently rising by about 3.7 millimeters per year. The Intergovernmental Panel on Climate Change (IPCC) projects that by 2100, global sea levels could rise between 0.38 meters (low-emission scenario) and 0.90 meters (high-emission scenario) (Figure 21). Even the most optimistic projections imply the disappearance of entire coastal communities and island nations.

Projected sea-level rise (2000-2100) under emission scenarios

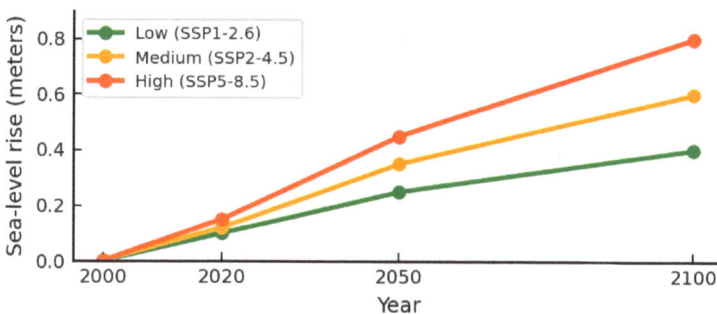

Figure 21. Projected sea level rise (2000–2100).
Source: IPCC AR6 (2021).

The implications are multiple:

- Human: Millions of people in island nations and low-lying coastal areas will face forced displacement, losing their homes, cultures, and local economies.

- Ecological: The shrinking polar ice threatens species such as polar bears, penguins, seals, and seabirds, disrupting entire food webs.

- Climatic: Less ice means less surface to reflect solar radiation, accelerating global warming and altering ocean currents and rainfall patterns.

However, while some communities prepare to lose everything, others see opportunity in the crisis. The melting of the Arctic is opening new shipping routes that cut travel distances between Asia and Europe by about 40% compared to the Suez Canal, and it is also unlocking access to oil, gas, and minerals once buried under the ice.

Major powers and corporations have begun competing for these territories, revealing a painful contrast: what for some is tragedy, for others is business.

This short-sighted vision—that places economic gain above life itself—is precisely what has brought humanity to this critical threshold. By turning crisis into an opportunity for profit, we accelerate planetary damage and deepen the very root of climate change.

The fate of the poles is not a distant matter. It is The Earth's thermometer, and a direct warning about the path we are taking as a species.

12. Insects

Over the past few decades, scientists have documented a dramatic decline in insect biomass. In protected areas of Germany, populations of flying insects dropped by roughly three-quarters in less than thirty years (Figure 22). This finding offers powerful evidence of how an essential group for ecosystems can collapse in a very short time.

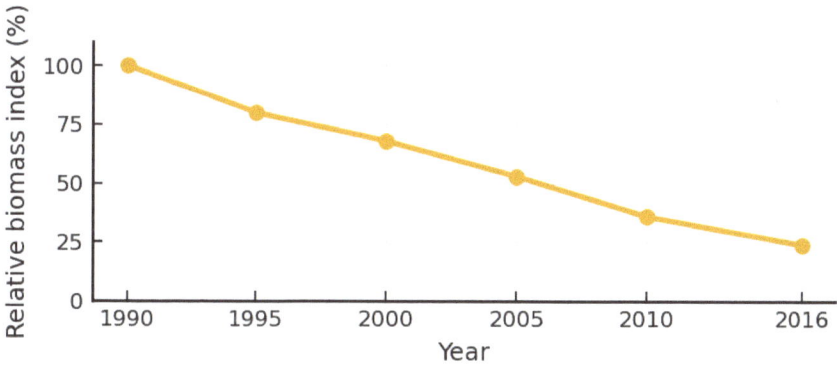

Decline of flying insect biomass protected areas in Germany (1989–2016)

Figure 22. Decline in flying insect biomass in areas of Germany (1989–2016). *Source: Hallmann et al. (2017).*

When we widen our view, we see that the problem is global. Different groups of insects—butterflies, bumblebees, beetles—have shown significant declines in their populations over the past forty years (Figure 23). This is not an isolated phenomenon, but a trend spanning continents and habitats.

Although the focus here is on insects, it's crucial to remember that other animal groups face equally severe risks. Amphibians are among the most threatened: more than 40% of species are at risk of extinction, their decline driven by habitat loss, disease, and climate change.

Among wild mammals, total biomass has dropped by more than 80% since historical times, and vertebrate populations overall have decreased by an average of 70% in the past fifty years.

These data make it clear that the crisis extends far beyond insects—it reflects a planetary unraveling of life.

Global decline of different insect groups (last 40 years)

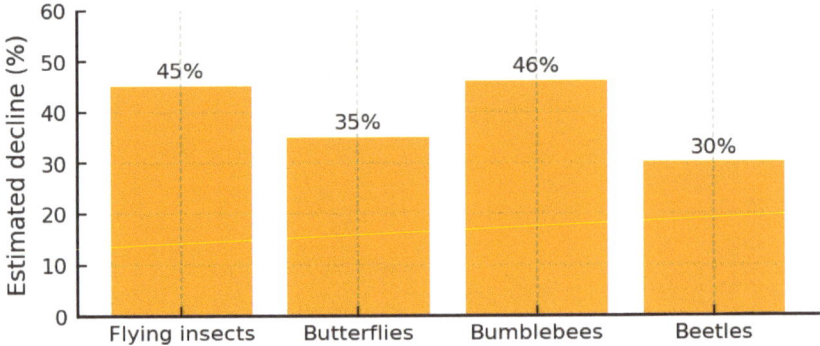

Note: "Flying insects" refers to all insect species capable of flight that have not been individually categorized as butterflies, bumblebees, or beetles.

Figure 23. Global decline of different insect groups.
Source: Sánchez-Bayo & Wyckhuys (2019); Hallmann et al. (2017).

The loss of insects is particularly critical because they form the very foundation of the web of life. They feed birds, amphibians, reptiles, and small mammals. They pollinate most crops and wild plants. They recycle nutrients by decomposing organic matter. They contribute to natural pest control and maintain soil fertility and aeration.

The main causes of decline include habitat destruction and fragmentation, the use of pesticides and pollutants, climate change, the introduction of invasive species, and the spread of diseases.

Despite the gravity of these trends, there are reasons for hope. Habitat restoration, reducing harmful chemicals, creating biological corridors, conservation programs, and growing social awareness can slow and even reverse these losses.

The message is clear: we still have time to act. By recognizing the importance of insects—and of all the forms of life that depend on them—we can protect the very base of the ecosystems on which our own existence also depends.

Recently, however, scientists have detected mosquito species capable of transmitting disease in places where they had never been seen before. The expansion of insects into regions where they could not previously survive is one of the clearest signs of global warming.

A striking recent example is Iceland: until this year, it was, along with Antarctica, one of the few places on Earth without mosquitoes. Yet in October 2025, the species Culiseta annulata—capable of surviving cold winters—was confirmed for the first time. This finding has been directly linked to the rapid temperature rise in the country, which is warming four times faster than the average of the Northern Hemisphere.

The arrival of mosquitoes in Iceland symbolizes how climate change reshapes not only landscapes but also the distribution of species, reconfiguring entire ecosystems in real time.

13. Mammals

About 10,000 years ago, at the dawn of agriculture, most of Earth's mammals were wild species. Humans represented only a small fraction within ecosystems regulated by natural forces.

Over the past millennia, the decline of mammals has been driven primarily by human activity: the conversion of forests, savannas, and wetlands into pastures, croplands, and cities; the fragmentation of landscapes by roads, dams, and fences that isolate populations; hunting and wildlife trade; the introduction of exotic species and pathogens and pollution—from pesticides, metals, plastics, noise, and light. More recently, global warming has shifted climates, altered food availability, and multiplied extreme events.

These pressures reduce populations, erode genetic diversity, and make species more vulnerable to local collapse.

Today, as shown in Figure 24, the biomass of mammals is dominated by humans and livestock, while the wild fraction has been drastically diminished. In just a few generations, we have reversed a balance that took Earth millions of years to build.

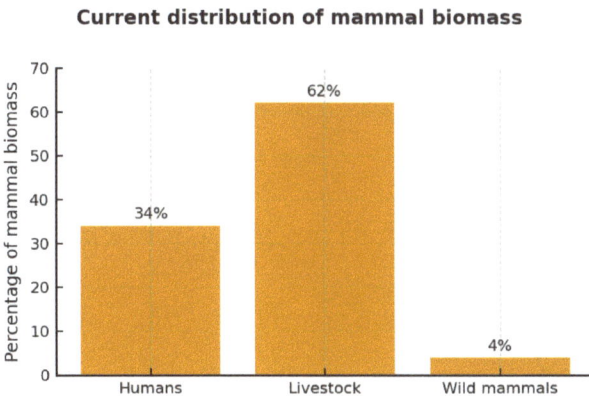

Figure 24. Distribution of mammal biomass.
Source: Bar-On et al. (2018).

This global pattern also has local expressions. In Puerto Rico, the Antillean manatee (*Trichechus manatus manatus*) faces threats directly linked to human activity. Figure 25 summarizes recorded mortalities between 1990 and 1995, classified as human-related, natural, or undetermined, based on published data for the Island.

These results reveal a profound shift in the composition of life: as human populations and livestock expand, wild mammals are being displaced.

Protecting them is not a luxury —it is essential to preserving the integrity of the ecosystems on which we all depend.

Causes of death of the Antillean manatee in Puerto Rico (1990-1995)
Grouped: human, natural and undetermined

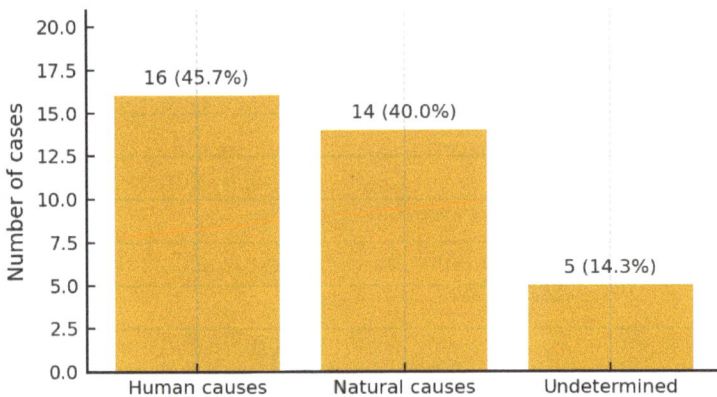

Figure 25. Causes of death of the Antillean manatee in Puerto Rico (1990–1995).
Source: Mignucci-Giannoni et al. (2000).

These numbers do not lie.

The oil we burn.
The pesticides we spread.
The forests we cut down.
The seas we empty.
The insects that vanish in silence.
The ice that melts at the poles.
The sea that rises.
The manatees —and countless other species— that die because of us.
The plastics we use for seconds.
The minerals we extract.
The invisible microbes we kill without knowing.
The waste we throw into the sky.
The wealth that concentrates in a few hands.

And the money we spend on war instead of restoring life —all point in the same direction: we are destroying the delicate layer that makes existence possible on this planet.

But there is also hope.

Because just as every human action contributes to the damage, every human action can also contribute to healing.

Changing how we use energy.

How we produce and consume food.

How we treat the forests, the oceans, the insects, the animals —and even the invisible microbes beneath our feet.

Choosing collaboration instead of competition, as nature itself does.

And demanding that the world's resources be directed toward healing our common home, not destroying it.

The Earth is presenting us with a clear account.

We can no longer say we didn't know.

And yet, there is still time—if we act together—to make those numbers change course and tell a different story: the story of a generation that chose to care for life, rather than end it.

References
The Earth's Mirror: Evidence And Data

Arava Institute for Environmental Studies. (2024). *Environmental–Humanitarian Impacts of War in Gaza*. [PDF report]. https://arava.org/wp-content/uploads/2024/06/Environmental-Humanitarian-Impacts-of-War-in-Gaza_reduced.pdf

Bar-On, Y. M., Phillips, R., & Milo, R. (2018). *The biomass distribution on Earth. Proceedings of the National Academy of Sciences*, 115(25), 6506–6511. https://doi.org/10.1073/pnas.1711842115

Benbrook, C. M. (2016). *Trends in glyphosate herbicide use in the United States and globally. Environmental Sciences Europe*, 28, 3. https://doi.org/10.1186/s12302-016-0070-0

Boucher, J., & Friot, D. (2017). *Primary microplastics in the oceans: A global evaluation of sources*. International Union for Conservation of Nature (IUCN). https://doi.org/10.2305/IUCN.CH.2017.01.en

BP. (2016). *Statistical Review of World Energy 2016*. BP. https://www.bp.com/statisticalreview

Consensus. (2020). *Number of satellites in orbit*. Consensus.

ClimateWorks Foundation. (2024). *Funding Trends 2023*. ClimateWorks Foundation. https://www.climateworks.org/report/funding-trends-2023

EIA. (2016). *International Energy Outlook 2016*. U.S. Energy Information Administration. https://www.eia.gov/outlooks/ieo/pdf/0484(2016).pdf

FAO. (2019). *FAOSTAT Pesticides Use Database*. Food and Agriculture Organization of the United Nations. https://www.fao.org/faostat

FAO. (2020). *Global Forest Resources Assessment 2020: Main Report*. Food and Agriculture Organization of the United Nations. https://doi.org/10.4060/ca9825en

FAO. (2021). *FAOSTAT Database*. Food and Agriculture Organization of the United Nations. https://www.fao.org/faostat

FAO. (2022). *The State of World Fisheries and Aquaculture 2022 (SOFIA 2022)*. Food and Agriculture Organization of the United Nations. https://doi.org/10.4060/cc0461en

FAOSTAT. (n.d.). *Fishery and Aquaculture Statistics*. Food and Agriculture Organization of the United Nations. https://www.fao.org/faostat

FAOSTAT. (n.d.). *Forestry Database.* Food and Agriculture Organization of the United Nations. https://www.fao.org/faostat

Forbes. (2024). *The World's Billionaires List 2000–2024. Forbes Magazine.* https://www.forbes.com/billionaires

Global Forest Watch. (2001–2023). *Global Forest Change Data.* World Resources Institute; University of Maryland. https://www.globalforestwatch.org

Global Tipping Points Project. (2025). *Global Tipping Points Report 2025.* University of Exeter, Global Systems Institute. https://global-tipping-points.org

Geyer, R., Jambeck, J. R., & Law, K. L. (2017). *Production, use, and fate of all plastic ever made. Science Advances, 3*(7), e1700782. https://doi.org/10.1126/sciadv.1700782

Hallmann, C. A., Sorg, M., Jongejans, E., Siepel, H., Hofland, N., Schwan, H., ... & de Kroon, H. (2017). *More than 75 percent decline over 27 years in total flying insect biomass in protected areas. PLOS ONE, 12*(10), e0185809. https://doi.org/10.1371/journal.pone.0185809

Hug, L. A., Baker, B. J., Anantharaman, K., Brown, C. T., Probst, A. J., Castelle, C. J., ... & Banfield, J. F. (2016). *A new view of the tree of life. Nature Microbiology, 1*(5), 16048. https://doi.org/10.1038/nmicrobiol.2016.48

IMBIE. (2018). *Mass balance of the Antarctic ice sheet from 1992 to 2017. Nature, 558*(7709), 219–222. https://doi.org/10.1038/s41586-018-0179-y

International Energy Agency (IEA). (2016). *Key World Energy Statistics 2016.* IEA. https://www.iea.org/reports/key-world-energy-statistics-2016

IPCC. (2021). *Summary for policymakers.* In *Climate Change 2021: The Physical Science Basis* (pp. 3–32). Contribution of Working Group I to the Sixth Assessment Report of the Intergovernmental Panel on Climate Change. Cambridge University Press. https://www.ipcc.ch/report/ar6/wg1/

ISAAA. (2020). *Global status of commercialized biotech/GM crops in 2019.* International Service for the Acquisition of Agri-biotech Applications (ISAAA). https://www.isaaa.org

McDowell, J. (2025). *Active satellites data.* Planet4589.org. https://planet4589.org

Mignucci-Giannoni, A. A., Beck, C. A., Montoya-Ospina, R. A., & Williams, E. H. (2000). *Mortality assessment of manatees (Trichechus manatus) in Puerto Rico, 1990–1995. Environmental Management, 25*(2), 189–198. https://doi.org/10.1007/s002679910018

NASA. (2024). *Arctic sea ice.* NASA Climate Change Vital Signs. https://climate.nasa.gov/vital-signs/arctic-sea-ice/

NASA. (2024). *Ice sheets.* NASA Climate Change Vital Signs. https://climate.nasa.gov/vital-signs/ice-sheets/

NASA GRACE/GRACE-FO. (2002–2023). *Gravity Recovery and Climate Experiment data.* National Aeronautics and Space Administration. https://grace.jpl.nasa.gov

National Snow and Ice Data Center (NSIDC). (2024). *Arctic sea ice minimum set (SSM/I and SSMIS).* University of Colorado Boulder. https://nsidc.org

NOAA / Puerto Rico Coral Reef Monitoring Program. (2005). *Coral reef condition: Stress factors and trends.* NOAA Coral Reef Conservation Program. https://www.ncei.noaa.gov/data/oceans/coris/library/NOAA/CRCP/NOS/OCM/Projects/198/NA13NOS4820009/Mercado2016_Monitoring_Fact_Sheet.pdf

OECD & FAO. (2021). *OECD-FAO Agricultural Outlook 2021–2030.* OECD Publishing. https://doi.org/10.1787/19428846-en

World Health Organization (WHO). (2020). *Pesticide residues in food.* https://www.who.int/news-room/fact-sheets/detail/pesticide-residues-in-food

Our World in Data. (2019). *Plastic production database.* University of Oxford. https://ourworldindata.org/plastic-pollution

Puerto Rico Department of Natural and Environmental Resources (DRNA). (n.d.). *Coral Reef Monitoring Program.* https://www.drna.pr.gov/coralpr/monitoreo/

Progressive Policy Institute. (2023). *Satellite Industry Trends Report.* PPI.

Sánchez-Bayo, F., & Wyckhuys, K. A. G. (2019). *Worldwide decline of the entomofauna: A review of its drivers. Biological Conservation,* 232, 8–27. https://doi.org/10.1016/j.biocon.2019.01.020

Shaheen, A., Dajani, R., Zinszer, K., Ashour, Y., & Abuzerr, S. (2024). *The war on the Gaza Strip and its consequences on global warming. Frontiers in Human Dynamics,* 6, 1463902. https://doi.org/10.3389/fhumd.2024.1463902

SIPRI. (2023). *Trends in world military expenditure 2023.* Stockholm International Peace Research Institute. https://sipri.org

Statista. (2024). *Number of billionaires worldwide from 2000 to 2024.* Statista. https://www.statista.com

The Guardian. (October 21, 2025). *Mosquitoes found in Iceland for the first time as climate crisis warms country.* https://www.theguardian.com/

environment/2025/oct/21/mosquitoes-found-iceland-first-time-climate-crisis-warms-country

Torres, A., Simoni, M., Keiding, J., & Østergaard, T. (2021). *Global sand trade: Current state, trends, and governance. Environmental Science & Policy, 116,* 93–101. https://doi.org/10.1016/j.envsci.2020.11.007

TS2 Space. (2025). *Number of satellites in orbit.* TS2 Space. https://ts2.space/en/satellite-count

UNEP. (2019). *Global Environment Outlook – GEO-6: Healthy Planet, Healthy People.* United Nations Environment Programme. https://www.unep.org/resources/global-environment-outlook-6

Union of Concerned Scientists. (2022). *UCS Satellite Database.* Union of Concerned Scientists. https://www.ucsusa.org/resources/satellite-database

United Nations. (2021). *Decade on Ecosystem Restoration: Financing Nature.* United Nations Environment Programme. https://www.decadeonrestoration.org

United States Geological Survey (USGS). (n.d.). *Mineral Commodity Summaries.* U.S. Department of the Interior. https://www.usgs.gov/centers/nmic/mineral-commodity-summaries

Whitman, W. B., Coleman, D. C., & Wiebe, W. J. (1998). *Prokaryotes: The unseen majority. Proceedings of the National Academy of Sciences, 95*(12), 6578–6583. https://doi.org/10.1073/pnas.95.12.6578

World Bank. (2020). *Minerals for Climate Action: The Mineral Intensity of the Clean Energy Transition.* World Bank. https://pubdocs.worldbank.org/en/961711588875536384/Minerals-for-Climate-Action-The-Mineral-Intensity-of-the-Clean-Energy-Transition.pdf

World Bank. (2022). *Nature's Frontiers: Achieving Sustainability through Investment in Nature.* World Bank. https://openknowledge.worldbank.org/handle/10986/37219

World Bank. (2023). *Military expenditure (% of GDP).* World Bank. https://data.worldbank.org/indicator/MS.MIL.XPND.GD.ZS

World Energy Council (WEC). (2016). *World Energy Resources 2016.* World Energy Council. https://www.worldenergy.org/publications/entry/world-energy-resources-2016

Paths That Already Exist

After the data we saw in The Earth's Mirror: Evidence And Data—the mirror reflecting the damage that Homo sapiens has inflicted upon the planet—this section opens the windows and lets the air in. It reveals something simple yet powerful: when people organize—communities, women, youth, Indigenous peoples—Earth responds. Each case presented here follows a different route, yet all share a common pattern: clear rights and rules, strong local institutions with a real voice, appropriate low-cost technologies, early benefits that motivate continuity, and constant evaluation to learn along the way. What follows are brief, accessible narratives written so that anyone can understand what existed before the problem, how it arose, who took action, what they did, and what is changing. The purpose is to inspire— and at the same time leave practical clues for starting small tomorrow and growing with persistence.

Communities Around the World

When we look for experiences in different countries, we are surprised by how many projects and people have spent years restoring, conserving, and caring for The Earth. Each initiative is a reminder that regeneration is possible when collective action is organized.

Initiatives in Puerto Rico

Here, too, in our archipelago, inspiring efforts exist. Local initiatives show how, with creativity and commitment, people can care for the land and sea, protect biodiversity, and open paths toward environmental justice from Caribbean contexts.

Enterprises that Care for The Earth

And it's not only communities or collectives: there are also entrepreneurs and businesses that have decided to change how they produce, sell, and relate to nature. From small local ventures to larger enterprises, many have joined this global movement—demonstrating that the economy can also be a force for hope. Some of these efforts are well-documented; others, not so much. There are even admirable initiatives that still lack a voice to share what they do. Therefore, what I present here is a sample: it is not meant to be exhaustive, nor to claim these are "the best" examples. This section is, above all, a seed—offering concrete ideas to look at one's own territory with new eyes, replicate what works, and keep searching. Learn from those who have already opened paths, often stumbling, until discovering effective ways to make it happen. All statements about microbiome restoration and ecological processes are supported by scientific literature listed in this section's references. What follows are not single recipes but diverse traces showing how life regenerates when we care for it.

Communities Around the World

The Netherlands · Giving the River Space

Throughout the 20th century, the Rhine and the Meuse were trapped behind ever-higher dikes. The order was clear: contain the water, straighten the meanders, gain land for cities and crops. But during the floods of 1993 and 1995, nature made its point: water rules. Thousands of people had to be evacuated; entire towns lived in suspense. Then came the question that changed everything: instead of building higher walls, what if we gave space back to the river? (Figure 1)

Ministries, provinces, and farmers redrew the map through patience and dialogue. Dikes were set back, levees lowered, relief channels built, and floodplains restored. There were compensations and, in some cases, relocations. It wasn't easy, but it was possible.

Figure 1. Floodplains along the Waal River near Bemmel (The Netherlands). *Source: Wikimedia Commons, public domain (CC0).*

Today, the results are visible: lower flood risk and more life along the banks—birds returning, wet meadows blooming, paths where people walk and watch the water. What was once a threat is now a living landscape.

Scotland · Rewetting the Peatlands

For decades, in the northern Highlands of Scotland, peatlands were drained and covered with conifer plantations. What had once been living sponges turned into dry fuel: soils losing carbon, landscapes prone to fire. Local communities, together with landowners and authorities, decided to reverse course. They began blocking drainage ditches, removing pines that didn't belong, and bringing water back to the soil. (Figure 2) The transformation is quiet but tangible: the damp silence of sphagnum moss growing again, the soil cooling in summer, the risk of fire fading, and cleaner water flowing downstream. This is not just about "reforesting." It's about rehydrating the land so life can do the rest. The essential change happens below the surface: the invisible return of microorganisms—the soil microbiome coming back to life, sustaining the peatland's natural cycle once again.

Figure 2. Rewetted peatland in northern Scotland.
Source: Wikimedia Commons (CC BY).

Spain (Semi-Arid Plateau) · Holding the Land with Living Agriculture

In the semi-arid southeast of Spain, almond and olive trees grew on bare soil. Mechanized tillage had stripped the slopes, and every storm tore away more earth—deep gullies, exhausted farms, discouraged farmers. But a group of farmers decided to change the story. They organized into networks such as AlVelAl, with technical support, and began transforming their land management. They sowed cover crops between tree rows, built micro-terraces and infiltration ditches to retain water, planted native hedges to break the wind, and integrated agroforestry—combining crops with useful trees.(Figure 3)

Gradually, the landscape began to green again in stripes. Soil organic matter increased, droughts hit less severely, and new sources of income appeared—organic almonds, rural tourism, value-added products. Where gullies once gaped, green seams now hold the soil. And the deepest change happens underfoot: regenerative practices awaken the soil's microbiome, restoring the invisible diversity of microorganisms that make the land fertile and alive.

Figure 3. Olive and almond groves on semi-arid hillsides in southeastern Spain. *Source: Wikimedia Commons, public domain (CC0).*

Portugal (Alentejo) · The Montado That Endures

The montado—the cork-oak savanna that covers much of Alentejo—had been weakening for years. Droughts, pests, overgrazing, and fires left deep marks on the landscape and on the cork economy.
But landowners, municipalities, and associations decided to act together. They adjusted livestock loads and adopted rotational grazing, carried out selective pruning to ease the trees' stress, protected young seedlings, and planted new cork oaks. They also managed underbrush, created firebreaks, built stone walls, and shaped contour terraces to slow runoff, as well as installing water points to ease the extreme summers. (Figure 4)

Over time, the montado began to recover its strength. Cork plantations reconnected with biodiversity—returning insects, birds, and pollinators that in turn support production. And beneath the surface, something deeper happens: this regenerative management reawakens the soil microbiome—the microbial biomass that sustains fertility, stores carbon, and gives the system resilience. Struck by many crises, the montado learned to endure by returning to its roots.

Romania (Bucharest) · A Wetland That Was Born on Its Own—and People Recognized It

On the outskirts of Bucharest, a hydraulic construction project was abandoned: a crater of earth and unfinished canals. It looked like a wasteland—but nature had other plans. Slowly, water accumulated and life moved in. Reeds, ducks, foxes, and hawks colonized the nameless hollow. (Figure 5)

Neighbors began to notice. Teachers, naturalists, and families joined forces to defend what they saw: it wasn't an empty lot—it was an urban wetland of value. They raised their voices in meetings, campaigns, and open letters.

After years of citizen activism, their pressure worked. The site was declared Văcărești Nature Park. Today it features boardwalks, park rangers, citizen science, and outdoor classrooms where children learn first-hand.

Figure 5. *Văcărești Nature Park* in Bucharest, Romania.
Source: Wikimedia Commons, photo by Tiia Monto (CC BY-SA 4.0).

In the middle of a capital dominated by traffic and concrete, thousands of people discover that nature can be fifteen minutes from home—if we make room for it. And beneath the surface, the story deepens: moisture regenerated the soil, restored microbial diversity, and reactivated invisible processes that gave the ecosystem its pulse back.

Germany–Czechia (Bavarian Forest–Šumava) · Letting the Forest Heal Itself

Strong storms and bark beetle outbreaks brought down vast pine forests along the German–Czech border. The obvious reaction seemed to be "clean up" and replant quickly. But the national parks proposed another path. (Figure 6)

They decided to wait—to protect what was fragile but leave the fallen trees where they lay. What many feared as abandonment turned into an opportunity. Beneath the fallen wood, birch, beech, and spruce sprouted in a mosaic. Birds, fungi, and insects returned—species that couldn't thrive in a uniform forest.

The lesson wasn't "do nothing" but "do just enough" so the forest can reorganize itself with greater diversity and resilience. Science supports it: fallen logs enrich the soil, increase microbial biomass, and add carbon and nutrients that feed beneficial fungi and bacteria.

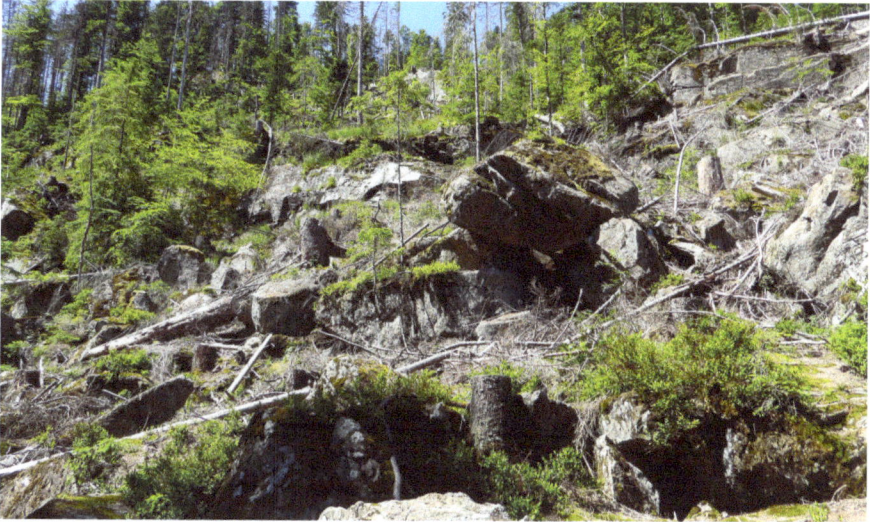

Figure 6. Regenerating forest in Bavarian Forest National Park.
Source: Wikimedia Commons, photo by Rosa-Maria Rinkl (CC BY 4.0).

Today, in this forest learning to heal itself, life regenerates from below—the soil microbiome revives and sustains a new diversity that is already spreading through the clearings.

China (Loess Plateau) · Healing an Entire Watershed

On the Loess Plateau, the hills—plowed to the brink—suffered knife-sharp rainstorms that stripped away fertile soil within minutes. Decades of overgrazing and slope tillage had left the land bare, rootless, eroding away.

But something changed. Farmers and local governments, with technical support, decided to act differently. They built terraces to slow water flow, banned grazing on fragile slopes, protected streambeds, and planted perennial vegetation. (Figure 7)

The same rain that once destroyed now infiltrates the earth and feeds crops. Grass covers the slopes again, erosion has decreased, and productivity has risen.

Healing a watershed is not a one-day job—it is a sustained pact with the land. And what happens beneath the surface is as valuable as what's seen above: restored vegetation regenerated the soil microbiome, reviving the diversity of microorganisms that once sustained its lost fertility.

Figure 7. Restoration in the Loess Plateau, China.
Source: Wikimedia Commons, photo by Vmenkov (CC BY-SA 3.0).

Niger (Sahel) · The Forest Was Hiding in the Stumps

In south-central Niger, the land looked tired. Since the mid-20th century, demand for firewood and charcoal, the expansion of farm plots, and laws that made trees state property had discouraged farmers: any sprout was pulled up "to avoid trouble." After the great Sahel droughts, fields were left bare and parched. But beneath the surface a secret remained: roots and stumps kept a "subterranean forest" alive, waiting for its chance. (Figure 8)

Figure 8. Stumps that seemed dead resprout in Niger.
Source: Wikimedia Commons (CC BY-SA 1.0).

Change arose within the villages themselves. Farmers in Maradi and Zinder began to ask: why do the stumps keep resprouting, even when we try to remove them? With technical support and local leadership, they flipped the logic: instead of yanking out the shoots, they protected them. They left four or five vigorous resprouts per stump and pruned the rest, shaping the trees like umbrellas. That way, millet and sorghum still got light, but the soil stayed cooler and more fertile under the shade.

In a few years, the bare lands filled with trees—no nurseries or plantations, just Farmer-Managed Natural Regeneration (FMNR). The soil firmed up, the wind raised less dust, there was fodder in the dry season, and firewood came from pruning, not felling. Beyond the visible, something essential re-ignited: the soil's invisible life. By dropping leaves, protecting roots, and avoiding poisons, the underground microbiome—bacteria, fungi, and mycorrhizae—reawakened, supporting fertility, moisture, and resilience.

Ethiopia (Tigray) · Hills That Rest to Bring Back Water

For years, Tigray's communal hills lived on the edge. Overgrazing, fuelwood cutting, and intense rains stripped the thin fertile layer and opened deep gullies. With few energy alternatives and fodder needs, every meter of soil was used to exhaustion. When the rains came, the bare ground retained nothing: water rushed down brown, and wells dried too soon.

The turn began with a simple yet hard idea: let the land rest. Exclosures were born—communal areas where grazing and wood extraction stopped to allow natural regeneration. Where seeds were lacking, assisted seeding was done; community guards were organized; and a "cut-and-carry" system was applied—the livestock stays out, but families harvest fodder without damaging the slope. In the most active gullies, small stone check dams slowed the water.

Within a few years, vegetation cover returned. The soil absorbed rain again, springs lasted longer, and both biomass and soil carbon began to increase. Beehives appeared on the edges and the community found enough fodder without harming the hillside.

An exclosure is not "fence and forget": it's living governance—simple rules, social monitoring, shared benefits. It turns hillside rest into water, honey, and life downstream. (Figure 9) Beyond the visible, the invisible revived too: highland studies show exclosures increase soil organic carbon and microbial biomass—clear signs of a reawakening soil microbiome that rescues the hidden life sustaining fertility.

Figure 9. Community hillside restoration in Ethiopia.
Source: Wikimedia Commons / treesftf (CC BY 2.0).

Rwanda · Restoring a Country of Hills, From the Soil Up

In Rwanda—the "land of a thousand hills"—soils were exhausted long before the great tragedy. Between 1960 and 1980, pressure to produce more pushed agriculture onto fragile slopes; fallows were cut short and conservation practices abandoned. With no rest, the land thinned: first rains carried off the topsoil; subsequent ones carved deep gullies. After 1994, mass resettlement intensified pressure: forests were felled to plant on unstable slopes and Gishwati and Mukura shrank to isolated patches.

The turning point was deciding to restore not only trees, but entire landscapes. The country mapped opportunities: agroforestry in some areas, silvopasture in others, forest regeneration where essential. A green fund financed projects and, by law, Gishwati and Mukura were joined into a national park with a buffer zone. Today, forest cores are expanding again. Hillsides hold water better and community agreements turn slogans into weekly shared tasks. (Figure 10)

Beneath the visible, a deeper transformation beats: forest and soil restoration have strengthened the underground microbiome. Research in Gishwati-Mukura shows improved soil biochemical properties and increased litter arthropods—clear signals of livelier, more functional soils that once again support life.

Figure 10. Hill restoration in Rwanda.
Source: Wikimedia Commons, photo by Antoinetorrens (CC BY-SA 3.0).

Kenya · Women Who Planted Water—and the Future

On eroded hills and in neighborhoods short of fuelwood and water, many women carried the landscape's weight on their backs. In the late 1970s, Wangari Maathai and neighborhood groups asked a simple, powerful question: what if we plant native trees near homes, schools, and riverbanks?

From that question, the Green Belt Movement was born. Community nurseries were organized, payments were made for each surviving seedling, and civic brigades protected forests and water sources. Every nursery was also a school: how to select local seeds, care for seedlings, water them in drought, and organize so what's planted grows strong. (Figure 11)

The results were deep and visible: millions of trees greened the hills, streams ran again in the dry season, soils stopped disintegrating at the first downpour, and thousands of women built networks with voice and power.

It wasn't just tree planting. It was reclaiming control of the landscape and the right to a healthy environment.

Figure 11. Wangari Maathai, founder of the Green Belt Movement.
Source: Wikimedia Commons (CC BY-SA 4.0).

Beneath that visible rebirth, something essential happened: the soil's invisible life reanimated. Restored vegetation increased organic carbon and moisture, multiplying microbial biomass. That underground life pulsed again, supporting more fertile, resilient soils.

Nepal · When the Forest Returned to Its Neighbors

In the 1970s and 80s, many Himalayan forests were dwindling. Scattered fuelwood and timber extraction, frequent fires, and centralized management that sidelined communities had turned them into land belonging to everyone—and to no one. With a growing population and soaring fuel demand, forests thinned before everyone's eyes.

The response was to give prominence to those who had always lived beside the forest. Community Forest User Groups (CFUGs) were recognized and given management rights. Each group demarcated its area, set quotas and seasons for use, organized patrols, and earmarked part of revenues for community funds. Women's participation and rotating responsibilities were promoted, while local technicians supported thinning, living fences, enrichment planting, and fire management. (Figure 12)

Over time, where there had been scrub there are now dense thickets and young stands. Fires decreased, fodder stabilized, and groups began to earn from both timber and non-timber products. Most importantly, the forest ceased to be nobody's and once again had caretakers—not to exclude, but to protect with shared rules.

Figure 12. Women of the Musahar community.
Photo: Jason Houston for USAID, Wikimedia Commons, public domain.

India (Rajasthan) · Water First, Then the Forest

In Rajasthan, after years of severe drought and rural exodus, seasonal rivers vanished and wells dropped deeper each season. It seemed the land was giving up. Villages and panchayats, guided by local leaders, decided to start with the essential: water. They rebuilt johads—small earthen dams—to capture monsoon runoff and recharge aquifers. (Figure 13)

Figure 13. Community johad in Laporiya, Rajasthan, India.
Photo: Amar Singh Kangarot, Wikimedia Commons, CC BY-SA 4.0.

With moisture came wetlands and riparian vegetation; crops stabilized again. And later—when water once more sustained life—planting was done where needed. The secret lay in multiplying many small works built and maintained by the community. Under that visible rebirth, a vital transformation unfolds: by returning water and allowing vegetation to return, soil regains organic matter and moisture. The microbiome blooms again—the microorganisms that sustain fertility and soil health.

Philippines · Opening the Tide So the Mangrove Can Heal

For years, mangrove "restoration" meant planting seedlings where tidal flow no longer reached. Roads and shrimp farms had blocked channels, leaving saline, stagnant soils incapable of sustaining the ecosystem's natural cycle.

Local communities, together with universities, reversed the order. First they reopened channels so the tide could recover its pulse. Only then did they plant where the system itself called for it. The result was immediate: seedling survival increased, fishing improved, and storm surges now meet a forest that slows them. The mangrove proved to be living engineering: if water flows, the mangrove knows what to do. (Figure 14)

Figure 14. Mangrove restoration in Guagua, Pampanga (Philippines). Photo: E911a, Wikimedia Commons, CC BY-SA 4.0.

Beneath that visible restoration, a quiet, profound transformation beats: restoring sea circulation and soil moisture also restored the microbiome. Microbial life that salinity and stagnation had stifled diversified again, sustaining ecosystem health from below.

China (Kubuqi) · Fixing Dunes, Opening the Future

In the Kubuqi Desert, dunes advanced like waves of sand. The landscape seemed in perpetual motion, swallowing roads, crops, and villages. Communities, authorities, and companies forged practical agreements: fix the sand with barriers and native species, harvest every drop of water, harness solar energy, and create local jobs tied to restoration.

Satellites confirmed what eyes already saw: stabilized dunes and green patches spreading through the desert. Where wind once beat without mercy, there is now work, horizon, and soil that can support life again. (Figure 15)

Beneath that evident rebirth, a silent transformation unfolds: revegetation increases organic carbon and nutrients, multiplying microbial diversity. Bacteria and fungi reweave soil's biological structure, driving ecological recovery from below.

Figure 15. Kubuqi Desert dunes with ancient structures (stupa). *Photo: Wikimedia Commons author, CC BY-SA 4.0.*

South Korea · Greening a Country, Village by Village

After the war, much of South Korea was deforested—bare mountains, exposed soils. Erosion carried off fertile earth and valleys dried. The government spurred a nationwide reforestation movement within everyone's reach. Local nurseries were organized, hands-on training offered, modest incentives for community tasks provided—and, above all, thousands were mobilized to own the project, village by village. It wasn't planting for planting's sake: erosion was controlled, the right species chosen, and care for each young tree ensured. Decades later, forests returned. Landslides decreased, valleys regained moisture, and the landscape became safer and more fertile. It was a national policy woven by people's hands.

Beneath that visible rebirth, a powerful transformation took place: reforestation increased organic matter, nutrients, and litter depth—creating conditions that reactivate microbial biomass. The soil microbiome flourished again, sustaining ecosystem fertility and vitality.

Figure 16. Bijarim Forest on Jeju Island, South Korea.
Photo: Project Manhattan, Wikimedia Commons (CC BY-SA 3.0).

Vietnam (Mekong Delta) · Mangroves as Natural Infrastructure

The delta's coastline was losing land each year, and storms hit harder. The response was to see mangroves not as secondary scenery but as living infrastructure.

Province by province, water regimes were managed, planting was done in suitable sites, and satellite monitoring guided corrections. Over time, green bands stretched along the coast—natural barriers that soften waves, support fisheries, and regain ground from the sea. Success wasn't a single big project but adaptive, sustained management. (Figure 17)

Under that green infrastructure, a quiet but essential transformation beats: restoring the natural hydrology also regenerates soil microbial communities. Diversified and active, they once again sustain the invisible functions that give resilience to coastal ecosystems.

Figure 17. Mangroves at sunset in the Mekong Delta, Vietnam.
Photo: Wikimedia Commons, public domain (CC0).

Mexico · When the Forest Became Communal Again

For decades, outside sawmills and logging concessions left many communities with degraded forests and no control over their own land. The toll was both ecological and social: trees felled without limits, eroded soils, and communities stripped of decision-making. From the 1970s and 80s onward—through legal reforms and local organization—ejidos and Indigenous communities reclaimed control. Community forest enterprises were born: assemblies decide how much, where, and how to harvest; local monitoring; reinvestment in nurseries, firebreaks, roads, and workshops; and, when appropriate, strict conservation areas.

In places like Oaxaca and Michoacán, the forest stopped being "someone else's loot" and became living heritage. Fires decreased, soils stabilized, local jobs emerged, and young people found reasons to stay. The forest returned not as a postcard, but as a rooted economy. (Figure 18)

Beneath that visible rebirth lies a deeper transformation: community forestry fosters organic matter and avoids degradation. In secondary or restored forests, the soil microbiome reorganizes within a few years, recovering diversity and function after regeneration.

Figure 18. Forest in Oaxaca, Mexico.
Photo: Reem Hajjar, OSU College of Forestry / Wikimedia Commons (CC BY-SA).

Brazil · Corridors in the Atlantic Forest and the Golden Lion Tamarin's Return

The Atlantic Forest—one of the planet's most diverse—was reduced to thousands of fragments after centuries of logging and agricultural expansion. What had been a continuous mantle turned into a mosaic— hard to sustain for species needing free movement. The Atlantic Forest Restoration Pact brought together organizations, municipalities, researchers, and producers to stitch the forest back together with green corridors. In Rio de Janeiro, the Golden Lion Tamarin Association combined science and community: restoring strategic patches, agreements with landowners, environmental education, and even yellow-fever vaccination campaigns to protect the species.

The result can be seen and heard: the forest is closing with green seams, and the golden lion tamarin—symbol of this fragmented ecosystem— has multiplied its population. (Figure 19)

Beneath the canopy, a transformation as powerful as it is quiet unfolds: the soil microbiome regenerates. In Atlantic Forest agroforestry systems, integrating shade-grown crops—like cacao—increases microbial diversity and activates key soil functions, favoring carbon and nitrogen cycling. That reborn microscopic life underpins biodiversity and the survival of emblematic species.

Figure 19. Golden lion tamarin (*Leontopithecus rosalia*) in its natural habitat.
Photo: Steve Wilson, Wikimedia Commons, CC BY-SA 2.5.

Costa Rica · Paying to Conserve... and the Forest Answered

In the 1970s and 80s, Costa Rica lost forests at an alarming rate. Thousands of hectares were felled each year for pasture and crops, and forest cover plummeted.

The response was simple and decisive: pay those who conserve. Through FONAFIFO, Payments for Environmental Services (PES) turned previously invisible values—water, soil, biodiversity—into income. Owners, co-ops, and communities signed contracts to conserve, manage, or reforest, while the State ensured stable funding via fuel taxes and international support.

The result was a historic turnaround: deforestation dropped sharply, forest cover recovered, and living from a living forest shifted from utopia to State policy. (Figure 20)

Beneath that visible rebirth, a quiet transformation occurs: in natural regeneration areas, microbial biomass rises as the forest matures and accumulates organic matter. It's the awakening of the invisible life that sustains ecosystem health and restores soil fertility.

Figure 20. Cloud forest in Monteverde, Costa Rica.
Photo: Cephas, Wikimedia Commons (CC BY-SA 4.0).

Colombia · The Day the Tide Returned to the Mangrove

On Caribbean and Pacific coasts, mangroves were suffocated by roads, shrimp ponds, and embankments. Seawater stopped flowing, soils became overly saline, and planted seedlings wouldn't survive. Coastal communities, universities, and authorities took a crucial step: reopen channels, restore the tidal pulse, and only then plant where necessary.

Change was immediate. With water's return, the mangrove began to heal on its own: seedlings thrived, crabs and fish came back, birds returned to nest, and villages felt storm surges less. (Figure 21) Beneath that visible rebirth, a quiet transformation occurred: restoring natural hydrology favored the recovery of microbial communities in sediments. That invisible diversity revitalized coastal soil functions—the foundation on which the whole ecosystem rests.

Figure 21. Mangroves in Ensenada de Utría, Pacific coast of Colombia.
Photo: Philipp Weigell, Wikimedia Commons (CC BY 3.0).

Peru (Amazon) · Shade-Grown Cacao to Heal the Frontier

In departments like San Martín, shifting cultivation and short-cycle crops exhausted soils and pushed into forest. Every few years, forest was felled and burned, leaving land weak and the agricultural frontier expanding.

The solution wasn't to expel people, but to change management. Shade-grown cacao, cultivated in agroforestry systems, was introduced. With steady technical assistance, secure land tenure, and access to quality-driven markets, families found a new path. Where slash-and-burn cycles once dominated, you now see treed plots that protect soil and provide more stable income. Cacao became a bridge—easing pressure on forests and restoring productive dignity for families. (Figure 22)

Beyond the visible, something crucial happened underground: in cacao agroforestry, microbial biomass and the diversity of bacteria, fungi, and archaea increased. The soil microbiome flourished again, sustaining long-term fertility and landscape health.

Figure 22. Shade-grown cacao agroforestry system in the Peruvian Amazon.
Photo: Roberto Fisher Cervantes, Wikimedia Commons (CC BY-SA 4.0).

Ecuador · "Socio Bosque": A Contract of Trust

In communities and families living in high-value ecosystems, reality cut both ways: poverty and deforestation. Need drove tree felling while biodiversity disappeared. Socio Bosque offered a clear, direct deal: periodic income in exchange for conserving forests and páramos. Agreements were long-term and consistently verified. The logic was social and ecological: stabilize local economies and recognize the value of those who care for the land.

The results speak for themselves. Independent evaluations show a notable drop in deforestation and community benefits that multiplied as financing stabilized: rangers, education funds, territorial pride, and new opportunities for youth. (Figure 23)

Underneath that visible achievement, an essential transformation takes place: by avoiding felling and maintaining cover, soils regain moisture and organic matter—offering refuge to thousands of microorganisms that reweave ecological functions from below, sustaining ecosystem health.

Figure 23. Native forest in Vilcabamba, Ecuador.
Photo: Marcelo Rosa Melo, Wikimedia Commons (CCo).

Cuba · Urban Agriculture in Hard Times

During the Special Period, the island ran out of fuel and cities lacked fresh food. The crisis reached every table. The response sprouted in lots, yards, and rooftops: organopónicos and urban gardens nourished by compost, biological control, and very little machinery. Co-ops and neighborhoods organized production and local sales, while technicians and agronomists offered simple, space-adapted solutions.

Havana discovered it could feed itself from within: local jobs, fresh vegetables, and living soils in the city itself. Urban agriculture wasn't a makeshift patch, but a school of resilience that transformed the city-land relationship. (Figure 24)

Behind that visible rebirth, something essential throbs: the awakening of the soil microbiome. Compost, crop rotation, and urban agroecological management restored organic matter and moisture, reactivating microbial biomass that sustains fertility even amid walls and asphalt.

Figure 24. Organopónico on the outskirts of Havana.
Photo: Arnoud Joris Maaswinkel, Wikimedia Commons (CC BY-SA 4.0).

Chile (Valdivian South) · Swapping Pines for Temperate Rainforest

On Chile's southern coast, pine and eucalyptus plantations had been displacing the Valdivian rainforest—one of the carbon-richest temperate forests on Earth. Soils were impoverished and native forest seemed doomed to scattered relics.

The Valdivian Coastal Reserve and local allies chose another path. They repurchased key parcels, protected what remained, and began converting exotic plantations into native forests—tepa, coigüe, ulmo. Local brigades were organized, ravines restored, and close work with neighboring communities began. (Figure 25)

Over time, native patches grew and started reconnecting mountains to sea. People reclaimed their forest—its fine rain, humid green, and springy, life-breathing soils.

Beneath this visible rebirth lies a key element: the soil microbiome. Native trees re-establish intimate partnerships with mycorrhizal fungi, essential to restore nitrogen and carbon cycles and return the ecosystem's lost vitality.

Figure 25. Valdivian Coastal Reserve in Los Ríos Region, Chile.
Photo: Natalia Reyes Escobar, Wikimedia Commons (CC BY-SA 4.0).

United States (Klamath River) · Removing Walls So the River Can Breathe

For over a century, reservoirs in the Klamath basin warmed the water, blocked sediment movement, and fueled cyanobacterial blooms that harmed the river and its fish.

Change rose from below. The Yurok and Karuk peoples—along with local allies and agencies—fought for decades for the same cause: free the river. In 2023–2024, four dams were dismantled and the channel ran free again.

Hundreds of miles of habitat opened to salmon. Restored flows lowered temperatures and harmful algal blooms began to recede. Today the tribes themselves, with technical support, continuously monitor periphyton, cyanotoxins, oxygen, and temperature to track the river's pulse. (Figure 26)

Most important, the invisible is shifting: as stagnant waters disappear and natural pulses return, the aquatic microbiome rebalances—fewer opportunistic cyanobacteria, more diverse communities recycling nutrients—the silent foundation of a river learning to breathe again.

Figure 26. Removal of Iron Gate Dam on the Klamath River. *Photo: International Rivers, courtesy; public reproduction authorized.*

Canada (British Columbia, Squamish Estuary) · Re-Opening the Tide's Pulse

For decades, a 1970s training berm cut off the Squamish estuary and left fry without refuge. The Squamish Nation, the Squamish River Watershed Society, the municipality, and local volunteers removed the barrier in phases: culverts were replaced, an 850-meter section of the spit was opened, and over 300 hectares of habitat were reconnected. The tide flowed in and out again, waters mixed, and estuary channels regained their "nursery" function for salmon.

With the water, the mud's microscopic life returned. Coastal-wetland literature shows that when a marsh's hydrology is restored, sediment microbial communities recover diversity and function (including denitrification), gradually approaching those of healthy marshes. The estuary's microbiome—the invisible "factory" that recycles nutrients—anchors the recovery. (Figure 27)

Figure 27. Aerial view of the Training Berm (Spit) in the Squamish River estuary.
Photo: Jesse Winter, The Narwhal, CC BY 4.0.

Initiatives in Puerto Rico

The Puerto Rican Crested Toad and the Puerto Rican Parrot

In the island's southwest, the Puerto Rican crested toad (Peltophryne lemur), Puerto Rico's only endemic toad, was declared locally extinct in the 1960s. Roads, urbanization, and invasive species destroyed its habitat and shrank its populations until it vanished from sight. The rescue began with captive breeding and mass releases of tadpoles into restored ponds. Today, organizations like Ciudadanos del Karso run breeding centers, and each season thousands of young toads return to forests and wetlands. (Figure 28)

A similar story unfolded with the Puerto Rican parrot (Amazona vittata), whose population fell to just thirteen individuals in the 1970s. Thanks to decades of scientific and community work—breeding, releases, and habitat restoration—the species has begun to recover. These efforts show that when a human community chooses to care for its most fragile fauna, life responds. The parrot's call and the crested toad's croak remind us that hope can also be bred, released, and multiplied. (Figure 29)

Figure 28. Puerto Rican crested toad (*Peltophryne lemur*).
Photo: Jan P. Zegarra, U.S. Fish & Wildlife Service, public domain.

The Puerto Rican parrot (*Amazona vittata*) survived in very few areas of El Yunque in the 1970s. Its recovery involved captive breeding, strategic releases, and installing artificial nests. Today it breeds in El Yunque, Río Abajo, and Maricao.

The shared lesson is clear: both the amphibian and the bird needed patience, consistency, and committed hands—not miracles. Applied science, genetic management, habitat restoration, and community participation combined to rescue two species on the brink.

Figure 29. Puerto Rican parrot (*Amazona vittata*) in its natural setting.
Photo: U.S. Fish and Wildlife Service, public domain.

Global Efforts

Across the world, organizations devote their work to defending The Earth and promoting sustainable alternatives for life. We can't list them all, but here are three representative examples of this global movement. They show that beyond local actions, there are international networks anyone can join to help care for our planet.

Jane Goodall Institute – Roots & Shoots

Founded by renowned primatologist and conservationist Jane Goodall, this program began in 1991 with a group of young people in Tanzania and now operates in over 60 countries. Roots & Shoots promotes projects led by children, youth, and communities seeking concrete solutions to protect the environment, animals, and people. Its core philosophy is simple but powerful: every action—however small—matters. Through school gardens, recycling campaigns, habitat restoration, and community projects, it inspires new generations to become Earth stewards.

La Vía Campesina

Founded in 1993, La Vía Campesina is an international alliance bringing together millions of peasants, family farmers, Indigenous peoples, rural workers, and social movements worldwide. It defends food sovereignty as a people's right: producing healthy, local food in harmony with nature and free from dictates of global markets. Its struggles include biodiversity defense, promotion of agroecology, and protection of the rights of those who work the land. It is a diverse, decentralized movement that weaves local experiences into a collective voice against the extractivist model.

Rainforest Foundation

Created in 1989, the Rainforest Foundation works with Indigenous and local communities to protect tropical forests and guarantee the rights of those who live in them. Projects include land demarcation, community forest monitoring, strengthening local leadership, and policy advocacy to stop deforestation. Active in Latin America, Africa, and Asia, it seeks a dual impact: conserving ecosystems critical to planetary climate stability while defending the human rights of the communities who depend on those forests.

Land and Environmental Defenders
Reverence, gratitude, and a contrast of values

The people who safeguard rivers, forests, mangroves, and territories do so out of ethical duty, without seeking privilege. Their commitment protects what gives us life—water, soil, biodiversity—and defends the values of care, cooperation, and the common good against the extractive logic.

What's happening?

Between 2012 and 2023, at least 2,106 land and environmental defenders were killed worldwide, according to Global Witness. In 2023, there were 196 killings, 85% of them in Latin America, with Colombia the deadliest country (79 victims), followed by Brazil (25), Mexico (18), and Honduras (18), according to Global Witness.

Why are they attacked?

Defenders are targeted where land and resources are disputed, especially in areas dominated by extractive economies. In 2023, mining was registered as the main sector associated with these killings, according to Global Witness.

Impunity and insufficient protection

Most of these crimes are neither investigated nor punished, feeding impunity. It's worth noting that the Escazú Agreement (2018) is the first regional treaty to include specific measures to protect those who defend the environment (Article 9), although its implementation still faces challenges, according to Global Witness.

Signs of hope

- Recognize and guarantee territorial and consultation rights, with early-warning mechanisms and community protection.
- Promote transparency and citizen participation in environmental decisions.
- Strengthen justice in the face of threats or attacks.
- Involve communities in building solutions.

Our word

We honor those who defended life and the planet. Their courage and values of cooperation and care uphold a sustainable and just future. Let us not forget: they build lasting wellbeing while others pursue fleeting gains.

Enterprises That Care for The Earth

We were sold convenience as modernity's grand prize: plastic bags, disposables, the instant. It's true they simplify life, but at the same time we're destroying ourselves. With every object we throw away, we not only dirty the land and the sea, we also feed a system that enriches a few at the expense of everyone else. True convenience would be reclaiming the pride of doing things well: fair, honorable work that sustains whole communities. Every small venture that repairs, reuses, or shares returns value to the local level: it keeps money circulating in our neighborhoods, towns, islands, or states instead of vanishing into distant corporations. And it gives us something deeper than income: the satisfaction of knowing we're caring for our Mother Earth. This isn't a passing fad. It's a return to older values—things that last, that are repaired, that are handed down. Because a society that discards easily ends up discarding itself.

Shoes with a new life
A broken shoe doesn't have to end up in the trash. A repair shop restores soles, stitching, and shine to what seemed lost. Each pair rescued saves money, reduces waste, and restores pride in manual craft.

Suitcases that travel again
Damaged wheels, handles, or zippers don't mean the end of a suitcase. Repair brings them back to the road while avoiding tons of plastic and fabric headed for the landfill.

A little clothing shop reinvented
Hems, buttons, embroidery, and creative mending can transform old clothing into unique pieces. What used to be waste becomes fashion with its own identity—threads that tell new stories.

Creative reuse center
Materials people no longer use—paper, fabric, buttons, wood—become treasures for schools, artisans, and workshops. A community reuse space sparks creativity and keeps tons of materials out of the trash.

Bike workshop
An abandoned bicycle can roll again with simple fixes. Neighborhood shops that rescue parts and teach people to repair their own bikes promote clean mobility, savings, and health.

Shared tools
Not everyone needs a drill or a sewing machine every day. A "library of things" lets neighbors borrow and share what's gathering dust in garages. That way we avoid over-buying and strengthen community cooperation.

Sharpening and upkeep of utensils
A knife or pair of scissors can last a lifetime if cared for. A small sharpening service extends the useful life of household and farm tools, saves money, and reminds us of the value of maintenance.

Neighborhood composting
Food scraps aren't trash: they're food for the earth. With community composters, we make soil that feeds gardens and urban farms, cutting methane in landfills and returning life to the ground.

Quality second-hand market
Clothing, furniture, and household goods can circulate many more times. An organized, trustworthy market that inspects and repairs what it receives offers good products at low prices and keeps value in the community.

Furniture that shines again
A worn sofa, a broken chair, a scratched table can regain strength and beauty through restoration or reupholstery. These workshops prevent unnecessary logging and create local jobs rich in craftsmanship.

Accessories and fashion from recycled materials
Bags from reclaimed fabrics, jewelry from reused metals, belts from recycled leather. Creativity turns leftovers into conscientious fashion—each piece carrying the mark of a cleaner planet.

Products for a plastic-free home
Cloth bags, beeswax wraps, plant-based sponges, bamboo brushes. Each one replaces a disposable, and together they make a big difference in reducing plastic in daily life.

Returnable containers and refills
Detergents, shampoos, and liquid soaps don't need to come in new bottles every time. A refill system with returnable containers saves packaging and fosters more conscious, local consumption.

Art with recycled materials
What others discard can become murals, sculptures, and urban furniture. Recycled art beautifies communities, creates identity, and shows that nothing is truly lost if there's imagination.

Recovered building materials
Doors, tiles, beams, metals, windows can be salvaged from demolitions and renovations. Reusing them in local projects reduces costs, saves resources, and keeps traditional building skills alive.

Electronics breathing again
Phones, laptops, and small appliances don't have to be disposable. Through repair, part replacement, and refurbished sales, technology can last longer and become more accessible—avoiding mountains of e-waste.

Regenerative urban agriculture
On rooftops, patios, or vacant lots, we can grow herbs, microgreens, or vegetables with compost, rainwater harvesting, and solar energy. Every urban garden is an oasis of life and a source of local, chemical-free, plastic-free food.

Collection of special wastes
Batteries, light bulbs, used oils, tires—items that shouldn't go to landfill can be picked up on community routes and taken to safe recycling. This protects soil and water health and creates responsible environmental jobs.

Handicrafts with plant fibers
Coconut husks, banana fibers, sugarcane bagasse, dried leaves—natural materials turned into baskets, handmade paper, or biodegradable utensils. Tradition meets innovation to cut plastic use.

Regenerative community tourism
Families and communities open their spaces to showcase gardens, forests, and living traditions, offering authentic experiences. Visitors learn, communities earn income, and nature becomes teacher and protagonist.

Closing

From Scotland to Puerto Rico, from Niger to Costa Rica, the stories in this section show that restoring The Earth also means restoring the invisible that sustains it: the soil microbiome. There, in that microscopic life that reappears when there is water, organic matter, and care, true resilience begins.

The lesson is clear: it's not about grand miracles, but about consistency, shared rules, and hands that sow hope. When the microbiome revives, The Earth responds—and with it, communities find a future.

But healing what's damaged isn't enough: we must also change how we live. For centuries, peoples knew how to reuse, recompose, repair, and share, because they understood everything was part of a cycle larger than themselves. We lost our way when we stepped off that path, handing power to corporations that squeeze The Earth's life to enrich a few.

Today, small enterprises that revalue the used, that repair instead of trash, that generate clean energy in their neighborhoods, and that return nutrients to the soil remind us there is another way. They are the human expression of the same natural recycling cycle that has sustained Earth for millions of years.

And that's the key: to re-embed ourselves in the planet's living cycle—not as intruders or owners, but as part of it. If we live this way—regenerating, reusing, recomposing—we won't just give Earth a breather; we'll build communities capable of sustaining themselves with dignity, allowing life to flourish again—for everyone.

References
Paths That Already Exist

Altieri, M. A., Companioni, N., Cañizares, K., Murphy, C., Rosset, P., Bourque, M., & Nicholls, C. I. (1999). *The greening of the "barrios": Urban agriculture for food security in Cuba. Agriculture and Human Values, 16*(2), 131–140.
Pioneering study documenting how urban agriculture in Cuba strengthened food security and community resilience in neighborhoods.

Antinori, C., & Bray, D. B. (2005). *Community forest enterprises as entrepreneurial firms: Economic and institutional perspectives from Mexico. World Development, 33*(9), 1529–1543.
Analysis of Mexican community forest enterprises (ejidos): economic performance, governance, and conditions that enable sustainable forest management.

Arévalo-Gardini, E., Arévalo, E., Baligar, V. C., & He, Z. L. (2015). *Changes in soil properties in long-term improved natural and traditional cacao agroforestry systems in the Peruvian Amazon. PLOS ONE, 10*(7), e0132147.
Evidence of improvements in soil properties under long-term cacao agroforestry systems in the Peruvian Amazon.

Arévalo-Gardini, E., Cantó, M., Alegre, J. C., et al. (2020). *Cacao agroforestry management systems effects on soil fungal diversity in the Peruvian Amazon. Ecological Indicators, 115*, 106404. https://doi.org/10.1016/j.ecolind.2020.106404
Shows how different cacao management practices under shade affect soil fungal diversity.

Arriagada, R. A., Ferraro, P. J., Sills, E. O., Pattanayak, S. K., & Cordero-Sancho, S. (2012). *Do payments for environmental services affect forest cover? Ecological Economics, 69*(11), 2116–2126.
Quasi-experimental evaluation of Costa Rica's payments for environmental services program; finds positive (though moderate) effects on forest cover.

Asociación AlVelAl. (2019). *Regenerative agriculture in the Steppe Highlands: Experiences and lessons.*
Landscape-scale transition case in southeastern Spain: soil restoration, almond/livestock management, and territorial cooperation.

Asociación Mico-Leão-Dourado. (2022). *Conservation of the golden lion tamarin in the Atlantic Forest.* Retrieved from https://www.micoleao.org.br
Flagship conservation program: habitat corridors, reintroductions, and community work to recover golden lion tamarin populations.

Asociación Parque Natural Văcărești. (2016). *Management plan of Parcul Natural Văcărești.*
Urban nature model in Bucharest: creation and participatory management of a metropolitan wetland as a natural park.

Baldrian, P., López-Mondéjar, R., & Kohout, P. (2023). *Forest microbiome and global change. Nature Reviews Microbiology, 21*(8), 487–501. https://doi.org/10.1038/s41579-023-00876-4
Review on the role of the forest microbiome in biogeochemical cycles and its relevance to resilience under global change.

BirdLife International. (2021). *Socio Bosque: Community conservation in Ecuador.* BirdLife. Retrieved from https://www.birdlife.org
Summary of a scheme that incentivizes communities and landowners to conserve native forests through agreements and conditional payments.

Bugalho, M. N., Caldeira, M. C., Pereira, J. S., Aronson, J., & Pausas, J. G. (2011). *Mediterranean cork oak savannas require human use to sustain biodiversity and ecosystem services. Frontiers in Ecology and the Environment, 9*(5), 278–286.
Demonstrates that dehesa/montado landscapes maintain biodiversity and ecosystem services thanks to well-managed traditional human use.

Bullock, R., Sheer, M., Nicol, R., & Wernick, A. (2021). *Indigenous-led salmon habitat restoration in the Pacific Northwest: Reviving ecosystems and communities. Ecological Applications, 31*(5), e02341.
Example of estuary and salmon habitat restoration led by Indigenous nations; ecological and sociocultural benefits.

Buyer, J. S., Baligar, V. C., He, Z. L., & Arévalo-Gardini, E. (2017). *Soil microbial communities under cacao agroforestry and cover crop systems in Peru. Applied Soil Ecology, 120,* 273–280.
Analyzes how cacao agroforestry and cover crop systems influence soil microbial diversity in the Peruvian Amazon.

Chen, Y., et al. (2024). *Sustainable vegetation thresholds on the Loess Plateau. Environmental Research: Ecology, 3,* 045001.
Recent study identifying critical vegetation cover thresholds for maintaining sustainability on China's Loess Plateau.

Daniels, A. E., Bagstad, K., Esposito, V., Moulaert, A., & Rodriguez, C. (2010). *Understanding the impacts of Costa Rica's payments for environmental services. Ecological Economics, 69*(11), 2116–2126.

Evaluation of Costa Rica's PES program; provides evidence on conservation and community welfare outcomes.

Departamento de Recursos Naturales y Ambientales de Puerto Rico (DRNA). (2024). *First Puerto Rican crested toad in the world born through in vitro fertilization*. San Juan: DRNA.
Historic news on assisted reproduction of the Puerto Rican crested toad (*Peltophryne lemur*), an endemic critically endangered species.

Departamento de Recursos Naturales y Ambientales de Puerto Rico (DRNA). (2025). *Programs and projects: Puerto Rican Crested Toad*. San Juan: DRNA.
Official update on conservation programs for the crested toad, including reintroductions and captive-breeding projects.

Duran, A. P., & Dickson-Hoyle, S. (2022). *Community forestry and Indigenous guardianship in Canada: Building resilience through collaborative governance*. Forest Policy and Economics, 137, 102705.
Explores how community forestry and Indigenous initiatives in Canada strengthen ecological and social resilience through collaborative governance.

Elion Group. (2021). *Kubuqi Desert restoration and green economy*. Elion Resources. Retrieved from http://www.elion.com.cn
China case: restoration of the Kubuqi Desert using revegetation techniques and large-scale green economy models.

Everard, M. (2016). *Community-based groundwater and ecosystem restoration in semi-arid north Rajasthan, India*. Ecosystem Services, 21, 20–30.
Example of aquifer and ecosystem restoration in India through community water management in semi-arid zones.

Food and Agriculture Organization of the United Nations (FAO). (various years). *Country profiles & forestry reports: Republic of Korea; Cuba (urban agriculture)*. FAO.
FAO reports compiling national experiences in urban agriculture (Cuba) and community forestry programs (South Korea).

Fondo Nacional de Financiamiento Forestal (FONAFIFO). (various years). *Evaluations of the Payment for Environmental Services (PES) Program*. Costa Rica.
Official documents analyzing the impact of Costa Rica's PES program on conservation and rural development.

Friess, D. A., et al. (2022). *Achieving ambitious mangrove restoration targets*. One Earth, 5(5), 456–460.
Global review of challenges and progress in large-scale mangrove restoration, linking science, policy, and communities.

Gebregergs, T., Tessema, Z. K., Solomon, N., & Birhane, E. (2019). *Carbon sequestration and soil restoration potential of grazing lands under exclosure management in a semi-arid environment of northern Ethiopia. Ecology and Evolution, 9*(11), 6468–6479.
Assesses carbon sequestration and soil restoration potential in grazing lands under exclosure management in Tigray, Ethiopia.

Green Belt Movement. (2022). *Our history and impact.* Nairobi, Kenya. Retrieved from https://www.greenbeltmovement.org
Organization founded by Wangari Maathai; has led the planting of over 50 million trees in Kenya, linking ecological restoration and social justice.

Huera-Lucero, T., Labrador-Moreno, J., Blanco-Salas, J., & Ruiz-Téllez, T. (2020). *A framework to incorporate biological soil quality indicators into assessing the sustainability of territories in the Ecuadorian Amazon. Sustainability, 12*(7), 3007.
Proposes biological soil quality indicators to assess sustainability in Ecuadorian Amazon territories.

Huera-Lucero, T., Torres, B., Bravo-Medina, C., García-Nogales, B., Vicente, L., & López-Piñeiro, A. (2025). *Comparative analysis of soil biological activity and macroinvertebrate diversity in Amazonian chakra agroforestry and tropical rainforests in Ecuador. Agriculture, 15*(8), 830.
Compares agroforestry chacras with tropical rainforests in Ecuador, highlighting positive effects on soil biological activity and macroinvertebrate diversity.

International Rivers. (2024). *Klamath River dam removal: Tribal leadership and ecological restoration.* Retrieved from https://www.internationalrivers.org
Documents dam removal on the Klamath River (USA), led by Indigenous tribes—one of the world's largest river restoration projects.

International Union for Conservation of Nature (IUCN). (2023). *Peltophryne lemur: Puerto Rican Crested Toad. The IUCN Red List of Threatened Species 2023.*
Official Red List entry for *Peltophryne lemur*, Puerto Rico's endemic critically endangered toad.

Jane Goodall Institute. (2024). *Roots & Shoots.* Retrieved from https://rootsandshoots.org
International program mobilizing youth and communities in over 60 countries with projects protecting people, animals, and the environment—showing that every action, however small, makes a difference.

La Vía Campesina. (2024). *International Peasant Movement.* Retrieved from https://viacampesina.org

Global movement founded in 1993, bringing together millions of peasants from all continents. Defends food sovereignty, promotes agroecology, and unites struggles for the rights of those who work the land.

Knapp, A. K., Blair, J. M., Briggs, J. M., Collins, S. L., Hartnett, D. C., Johnson, L. C., & Towne, E. G. (1999). *The keystone role of bison in North American tallgrass prairie restoration. BioScience, 49*(1), 39–50.
Shows how bison act as a keystone species in restoring North American tallgrass prairies.

Lavergne, C., Cabrol, L., Cuadros-Orellana, S., Quinteros-Urquieta, C., Stoll, A., Yáñez, C., Tapia, J., Orlando, J., & Rojas, C. (2024). *Rising awareness to improve conservation of microorganisms in terrestrial ecosystems: Advances and future directions in soil microbial diversity from Chile and the Antarctic Peninsula. Frontiers in Environmental Science, 12,* 1326158.
Scientific review on the importance of conserving microbial diversity in Chilean and Antarctic soils, aimed at ecological restoration.

Lewis, R. R. (2005). *Ecological engineering for successful restoration of mangroves. Ecological Engineering, 24*(4), 403–418.
Describes ecological engineering principles applied to successful mangrove restoration.

Lynn, K., & Gerlitz, W. (2020). *Tribal adaptation and resilience in the United States: Lessons from community-led restoration. Climatic Change, 160*(1), 55–72.
Analyzes community-led restoration in U.S. tribal communities as examples of climate change adaptation.

Maathai, W. (2004). *The Green Belt Movement: Sharing the approach and the experience.* Lantern Books.
Book in which Wangari Maathai shares the philosophy and achievements of the Green Belt Movement, integrating environmental conservation and women's empowerment.

Mekuria, W., Veldkamp, E., Corre, M. D., Haile, M., & Nyssen, J. (2011). *Restoration of ecosystem carbon stocks following exclosures. Soil Science Society of America Journal, 75*(1), 246–256.
Demonstrates how exclosure areas in Ethiopia help restore ecosystem carbon stocks.

Merloti, L. F., Mendes, L. W., Tsai, S. M., Geisen, S. A., & van der Putten, W. H. (2022). *The role of soil microbiome for ecosystem restoration in the Atlantic Forest. Proceedings of the 18th International Symposium on Microbial Ecology.*
Discusses the relevance of the soil microbiome in Atlantic Forest restoration, presenting recent advances in applied microbial ecology.

Ministerio del Ambiente, Agua y Transición Ecológica. (2021). *Socio Bosque Program: Results and impacts.* Quito, Ecuador. Retrieved from https://www.ambiente.gob.ec
Official summary of Ecuador's Socio Bosque program, combining economic incentives and community conservation in Indigenous and rural territories.

Müller, J., Bußler, H., & Blaschke, M. (2010). *Diverse response of arthropods to bark-beetle outbreak in Bavarian Forest National Park. Forest Ecology and Management, 259*(3), 300–309.
Shows how different arthropod groups respond to bark beetle outbreaks, offering lessons on temperate forest resilience.

The Narwhal. (2022, September 17). *Inside a 50-year journey to reopen the "lungs" of the Squamish River. The Narwhal.* Retrieved from https://thenarwhal.ca/squamish-nation-estuary-restoration
Journalistic chronicle of half a century of efforts to restore the Squamish River estuary in Canada, led by the Squamish Nation.

NatureScot. (2021). *Peatland ACTION: Progress and lessons from peatland restoration in Scotland.*
Report on peatland restoration in Scotland, highlighting achievements in carbon mitigation and biodiversity.

Pacto pela Restauração da Mata Atlântica. (various years). *Restoration reports and goals.*
Compilation of goals and results from Brazil's largest forest restoration pact, involving multiple public and private actors.

Pagiola, S. (2008). *Payments for environmental services in Costa Rica. Ecological Economics, 65*(4), 712–724.
Case study analyzing how payments for environmental services have helped conserve forests and generate local benefits in Costa Rica.

Premat, A. (2005). *Moving between the plan and the ground: Shifting perspectives on urban agriculture in Havana, Cuba.* In L. J. A. Mougeot (Ed.), *Agropolis* (pp. 171–204). IDRC.
Analysis of how urban agriculture in Havana has evolved between institutional planning and local practice, with implications for urban food sovereignty.

Reij, C., Tappan, G., & Smale, M. (2009). *Re-greening the Sahel: Farmer-led innovation in Niger.* IFPRI/USGS.
Documents how farmers in Niger have restored degraded lands at large scale through simple water and tree management practices.

Reserva Costera Valdiviana – The Nature Conservancy Chile. (2020). *Conservation and restoration of the Valdivian rainforest.* Retrieved from https://www.nature.org
Institutional report on conservation and restoration efforts in Chile's

Valdivian rainforest, including endemic species and local community work.

Rijkswaterstaat. (2015). *Room for the River: Programme Summary.* Ministry of Infrastructure and the Environment, Netherlands. Innovative Dutch program that returned space to rivers to reduce flood risks while combining safety and ecological restoration.

Rivas, Y., Stoll, A., & Godoy, R. (2023). *Microbial community and enzyme activity of forest soils under native and exotic species in southern Chile. Forests, 14*(6), 1221. Analyzes how native and exotic forests influence soil microbial communities and enzymatic activity in Chile.

Smith, A. C., et al. (2023). *Community forest management led to rapid local forest cover gains. Land Use Policy, 132,* 106021. Recent evidence that community forest management can produce rapid increases in local forest cover.

Somarriba, E., et al. (various years). *Cacao agroforestry in the Amazon.* CATIE. Compilation of cacao agroforestry experiences across Latin America, highlighting productive and environmental benefits.

Squamish River Watershed Society. (2023). *Central Estuary Restoration Project updates.* Retrieved from https://www.squamishwatershed.com Community report on the central estuary restoration project of the Squamish River, Canada.

The Nature Conservancy (TNC) – Chile. (various years). *Valdivian Coastal Reserve: Conservation and restoration reports.* Periodic TNC reports on progress and challenges in Valdivian rainforest conservation.

United Nations Convention to Combat Desertification (UNCCD). (2015). *Review of the Kubuqi Ecological Restoration Project.* International evaluation of ecological restoration in China's Kubuqi Desert, considered a global model.

United States Fish and Wildlife Service (USFWS), & DRNA. (2019). *Recovery Plan for the Puerto Rican Parrot (Amazona vittata).* USFWS. Official recovery plan for the Puerto Rican parrot, an endangered species, developed in collaboration with Puerto Rico's DRNA.

United States Geological Survey (USGS). (various years). *Farmer-Managed Natural Regeneration (FMNR) in Niger.* USGS. Technical documentation on farmer-managed natural regeneration in Niger, a successful case of community-based restoration.

USAID Biodiversity & Forestry. (2017). *Community forestry in Nepal: Measuring impact.* USAID. Retrieved from https://www.usaid.gov

Report on the impact of community forestry in Nepal, highlighting social and environmental results.

Valarezo Torres, G. E., Ochoa-Cueva, P., & others. (2021). *Soil quality/ health indicators in a disturbed ecosystem in southern Ecuador. Soil Science Annals, 72*(1), 17–26.
Evaluates soil quality indicators in disturbed ecosystems of southern Ecuador, with implications for restoration.

Wangari Maathai Institute for Peace and Environmental Studies. (2021). *The legacy of the Green Belt Movement.* University of Nairobi.
Summarizes the legacy of the Green Belt Movement and its impact on conservation, education, and community empowerment in Kenya.

World Bank. (2007). *Restoring China's Loess Plateau.* World Bank.
Flagship case of soil restoration in China's Loess Plateau that transformed agricultural productivity and reduced poverty.

World Future Council. (2019). *FMNR: Policy brief.*
Policy recommendation on farmer-managed natural regeneration as a large-scale restoration solution.

World Resources Institute (WRI). (2016). *Forest and landscape restoration in Rwanda: Gishwati-Mukura case study.* WRI. Retrieved from https://www.wri.org
Case study of forest restoration in Rwanda combining ecological and social benefits in the Gishwati-Mukura landscape.

WWF Mexico. (2018). *Community forest enterprises: The Mexican model.* World Wide Fund for Nature – Mexico.
Describes the model of community forest enterprises as a strategy for conservation and rural development in Mexico.

WWF Peru. (2020). *Shade-grown cacao: A sustainable alternative in the Peruvian Amazon.* World Wide Fund for Nature – Peru.
Example of shade-grown cacao production in Peru combining profitability and sustainability.

WWF Philippines. (2019). *Mangrove restoration in the Philippines: Community approaches.* World Wide Fund for Nature – Philippines.
Report on mangrove restoration in the Philippines highlighting effective community approaches.

Acknowledgments

So many people have touched my life, in such diverse ways —each one equally important.

I am grateful to every person who took the time and shared their knowledge to answer my questions, clarify my doubts, and guide me toward directions that led to understanding. And above all, to those who made me think—and rethink—my ideas by asking the difficult questions.

Some did so for only a few minutes; others walked beside me for years. But the seed they left—whether small or large—took root and bore fruit.

I owe special thanks to Dr. Betzaida Ortiz Carrión, for her careful review of the manuscript and her constructive criticism.

This book also belongs to all of you.

About The Author

Josefina Arce, Ph.D.

Josefina Arce is Professor Emerita of Chemistry at the University of Puerto Rico, where for more than three decades she led visionary projects that transformed science education.

Recognized as a pioneer of the "science for all" movement, she has mentored generations of scientists and educators, and her collaborations with the National Science Foundation marked a turning point in the country's approach to scientific training.

In 2017, she faced Hurricane María on her 75-acre farm in Santa Isabel, a semi-arid region in southern Puerto Rico. After losing her entire harvest, she spent eleven months without electricity or running water. During the first week, she was completely cut off from the outside world—no cell phone, no internet, no radio, no television— knowing only what she could see with her own eyes.

That experience of isolation, loss, and resilience deepened her commitment to regenerative agriculture and food sovereignty.

Today, from Finca Atabey, Josefina brings together science, art, and agriculture to inspire new generations to discover their place in the universe and to care for The Earth as our only home.

Contact
Email: josefinaarce@gmail.com
Website: www.agroempresasatabey.com
Facebook: facebook.com/fincaatabey

After The Last Page...

This book ends,
but what it awakened in you doesn't have to fade.
If something inside you stirred—a certainty, a question, an urgency—
you will not be alone.

All around the world, many of us are remembering
that caring for life is not just an idea:
it is a daily practice born of the body, the heart, and understanding.

That is why we have created a living, intimate, ongoing space—
a place to keep deepening what this book has opened:
real conversations, learning that transforms us,
and concrete actions that nurture life.

There we share ideas, resources, questions, and stories.
We accompany one another.
We unlearn and learn together.
We look again at what is essential—with new eyes.
And we follow the pulse of The Earth.

If you feel the call, join the Facebook group:
Red Viva de la Tierra

(Living Earth Network)

This is not a full stop.
It is a root extending outward—
a minga that has just begun.

Minga: *collective work done with joy, where what is cultivated belongs to everyone.*

www.ingramcontent.com/pod-product-compliance
Lightning Source LLC
Chambersburg PA
CBHW040932050426
42334CB00047B/111